Passive Butterworth Filter Cookbook

Stefan Hollos and J. Richard Hollos

Passive Butterworth Filter Cookbook
by Stefan Hollos and J. Richard Hollos

Copyright ©2021 by Exstrom Laboratories LLC

Abrazol Publishing
an imprint of Exstrom Laboratories LLC
662 Nelson Park Drive, Longmont, CO 80503-7674 U.S.A.

Publisher's Cataloging in Publication Data
Hollos, Stefan
Passive Butterworth Filter Cookbook / by Stefan Hollos and J. Richard Hollos
p. cm.
ISBN: 978-1-887187-42-8
Library of Congress Control Number: 2021942490
1. Electric filters–Design and construction–Handbooks, manuals, etc. 2. Electric filters–Design
I. Title. II. Hollos, Stefan.
TK7872.F5 .H65567 2021
621.381 HOL

Contents

PREFACE

This book should allow anyone with basic electronics skills to quickly design a passive Butterworth filter. All possible low pass and high pass filters up to tenth order, and all possible band pass and band stop filters up to eighth order are covered. This means schematics and component values for these filters are given along with formulas for scaling the values to the particular frequency the filter must operate at.

For each filter type and order there are four variations. Variation 1 is for a finite source resistance and infinite or very high termination resistance. Variation 2 is for an ideal zero resistance source and a finite termination resistance. Variations 3 and 4 are for the balanced case of equal source and termination resistance.

There is a design example for each of the filter types that shows how to scale the component values. A Spice simulation file for the design is given along with the frequency response. Running a simulation on a filter is useful for looking at the effect that non ideal components have on the frequency response. Some inductors for example may have significant series resistance which can be included in the simulation.

If you want band pass or band stop filters higher than eighth order the book explains how to use a low pass filter to construct them. So the information is there to construct filters up to twentieth order but we don't recommend trying to do that. The non ideal nature of the components makes it hard to get the expected extra performance.

The alternative to a passive Butterworth filter is of course an active Butterworth filter but active filters do not perform well at frequencies above 100 kHz. The passive filters presented in this book are usually the only practical choice for frequencies from 1 MHz up to about 1 GHz. They are also the only choice in situations where the power required for active filters is not available such as in audio crossover networks inside speaker enclosures.

In some cases only a simple first or second order filter will do the trick so the book starts out discussing these filters. They require at most one inductor which makes them easy to design and build. Even if you are sure you need a higher order filter it is a good idea to read over this material for general insights into filter behavior we think it provides.

You can find more information on signal processing and other engineering related topics at our website:

http://www.exstrom.com/exstrom/pande/index.html

Information on digital Butterworth filters can be found in our book "Recursive Digital Filters: A Concise Guide". The website for that book is http://www.abrazol.com/books/filter1/

Finally, the website for the book you are now reading is http://www.abrazol.com/books/filter2/ where we will post related resources.

FIRST ORDER FILTERS

With a single capacitor or inductor and a resistor you can create a first order filter. Figure 1 shows a first order low pass filter composed of one resistor and one capacitor.

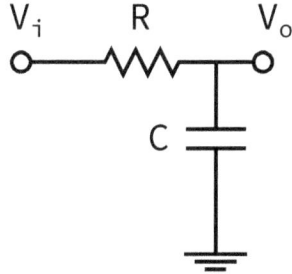

Figure 1: First order low pass filter using a capacitor.

The input voltage is applied to the terminal marked V_i and the output voltage is taken from the terminal marked V_o. We will assume that whatever the output terminal is connected to has infinite input impedance so that it does not load the filter. In that case when V_i is a DC voltage the capacitor behaves like an open circuit and $V_o = V_i$. As the frequency of V_i increases the impedance of the capacitor decreases and so does V_o since the filter is really just a simple voltage divider. We will call the frequency at which $V_o = V_i/\sqrt{2}$ the cutoff frequency for the filter. It may also sometimes be called the 3 dB or half power frequency. For this filter the cutoff frequency is $\omega_0 = 1/RC$ where $\omega_0 = 2\pi f_0$ and f_0 is the frequency in Hz. Let $\omega = 2\pi f$ be the frequency of V_i and $x = \omega/\omega_0$ then the overall frequency response of this filter is given by the function

$$g(x) = \frac{1}{\sqrt{1 + x^2}} \tag{1}$$

A plot of this function is shown in figure 2.

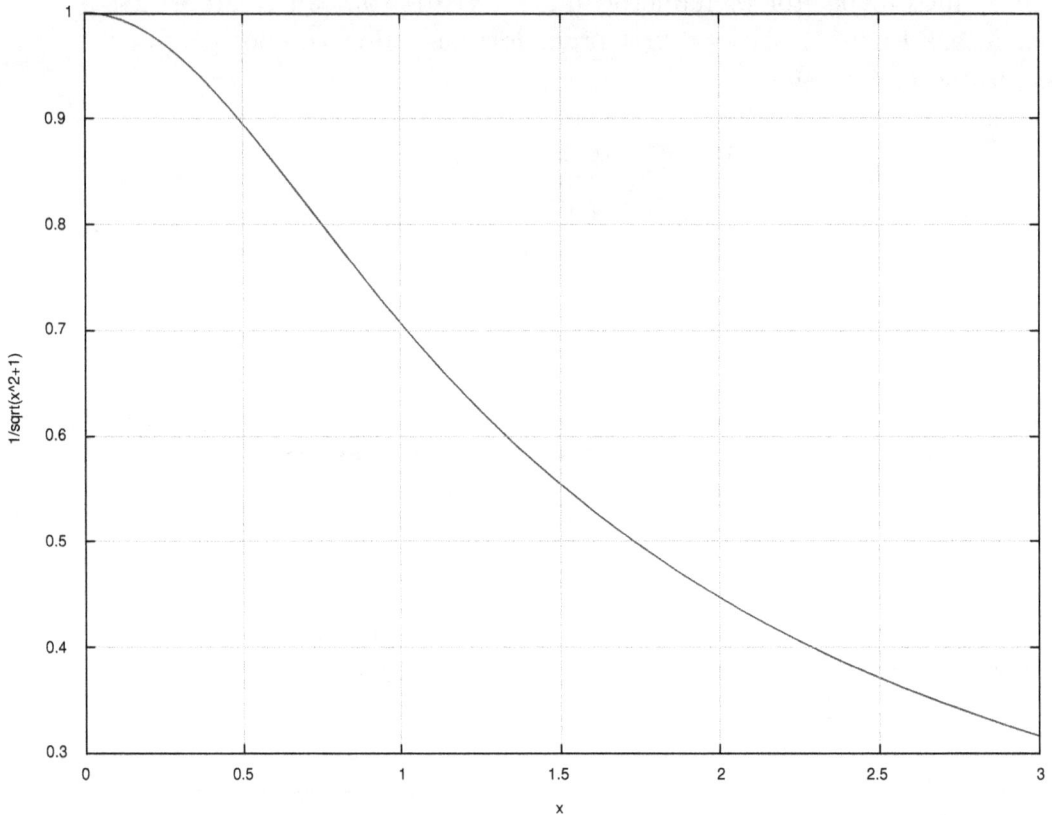

Figure 2: Frequency response of first order low pass filter.

Note that when $x = 1$ or $\omega = \omega_0$ then $g(x) = 1/\sqrt{2}$. Figure 3 shows the phase response of the filter.

This is the phase difference between V_i and V_o as a function of frequency. The phase response is $\theta(x) = -\arctan(x)$. To design this filter given R

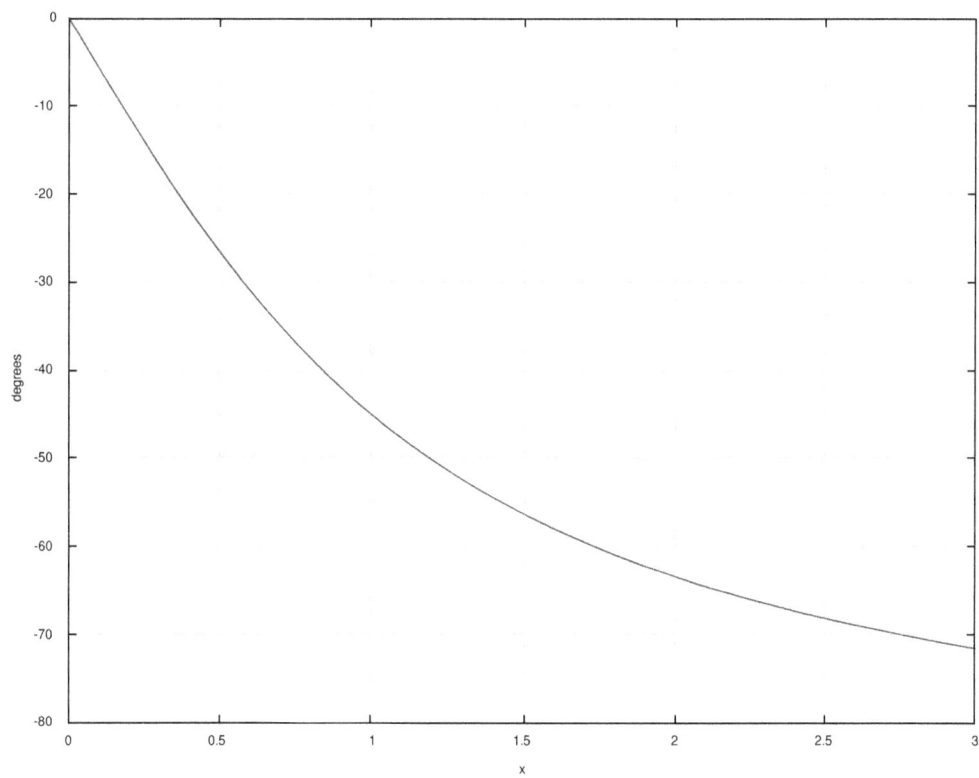

Figure 3: Phase response of first order low pass filter.

and ω_0 let $C = 1/R\omega_0$. The derivation of these formulas is as follows.

$$G(s) = \frac{V_0}{V_i} = \frac{1}{RCs + 1} \tag{2}$$

$$G(j\omega) = \frac{1}{j\omega RC + 1} = \frac{1}{1 + j\omega/\omega_0} = \frac{1}{1 + jx} = \frac{e^{-j\arctan(x)}}{\sqrt{1 + x^2}} \tag{3}$$

$$g(x) = |G(j\omega)| = \frac{1}{\sqrt{1 + x^2}} \tag{4}$$

$$\theta(x) = \arg(G(j\omega)) = -\arctan(x) \tag{5}$$

Figure 4 shows a first order low pass filter composed of one resistor and one inductor.

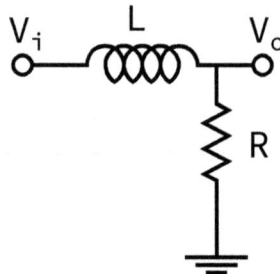

Figure 4: First order low pass filter using an inductor.

The frequency response is identical to the one for the RC low pass filter given above. The cutoff frequency in this case is given by $\omega_0 = R/L$. To design this filter given R and ω_0 let $L = R/\omega_0$.

Figure 5 shows a first order high pass filter composed of one resistor and one capacitor.

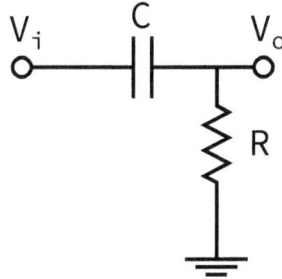

Figure 5: First order high pass filter using a capacitor.

At DC the capacitor looks like an open circuit and $V_0 = 0$. As the frequency increases the capacitor impedance drops and V_0 increases. At $\omega_0 = 1/RC$, $V_0 = V_i/\sqrt{2}$, and V_0 continues to increase as the frequency gets larger than ω_0 until $V_0 = V_i$ at $\omega = \infty$. The frequency and phase response for the filter are

$$g(x) = \frac{x}{\sqrt{1 + x^2}} \tag{6}$$

$$\theta(x) = \frac{\pi}{2} - \arctan(x) \tag{7}$$

The plots are shown in figures 6 and 7.

The derivation of these formulas is as follows.

$$G(s) = \frac{V_0}{V_i} = \frac{RCs}{RCs + 1} \tag{8}$$

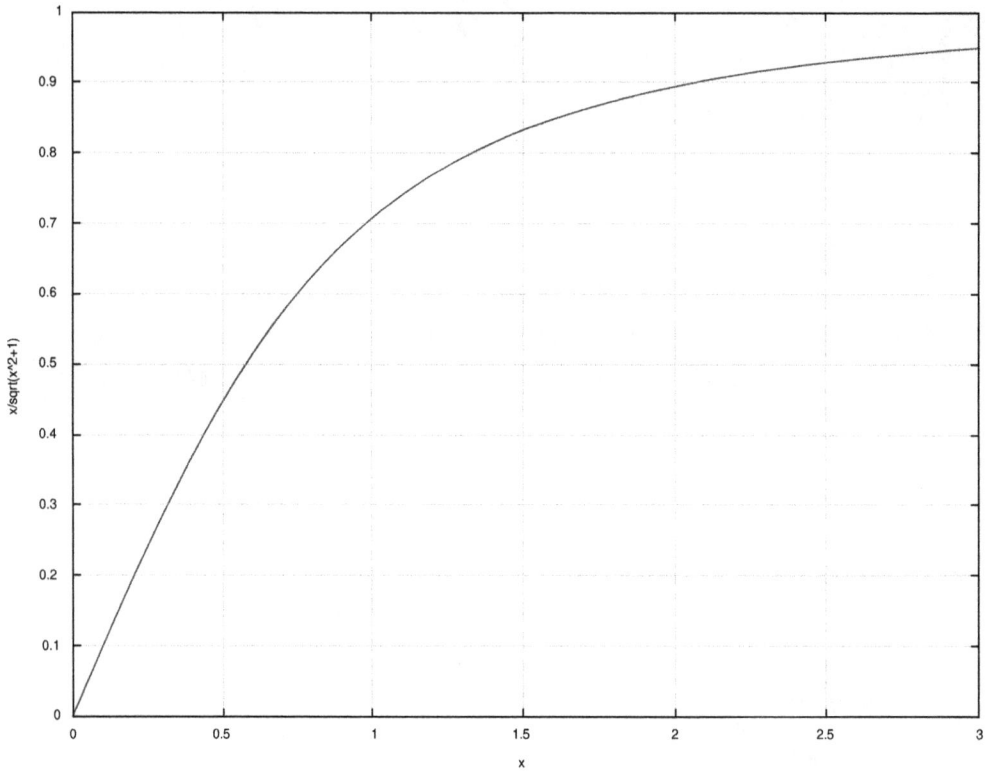

Figure 6: Frequency response of first order high pass filter.

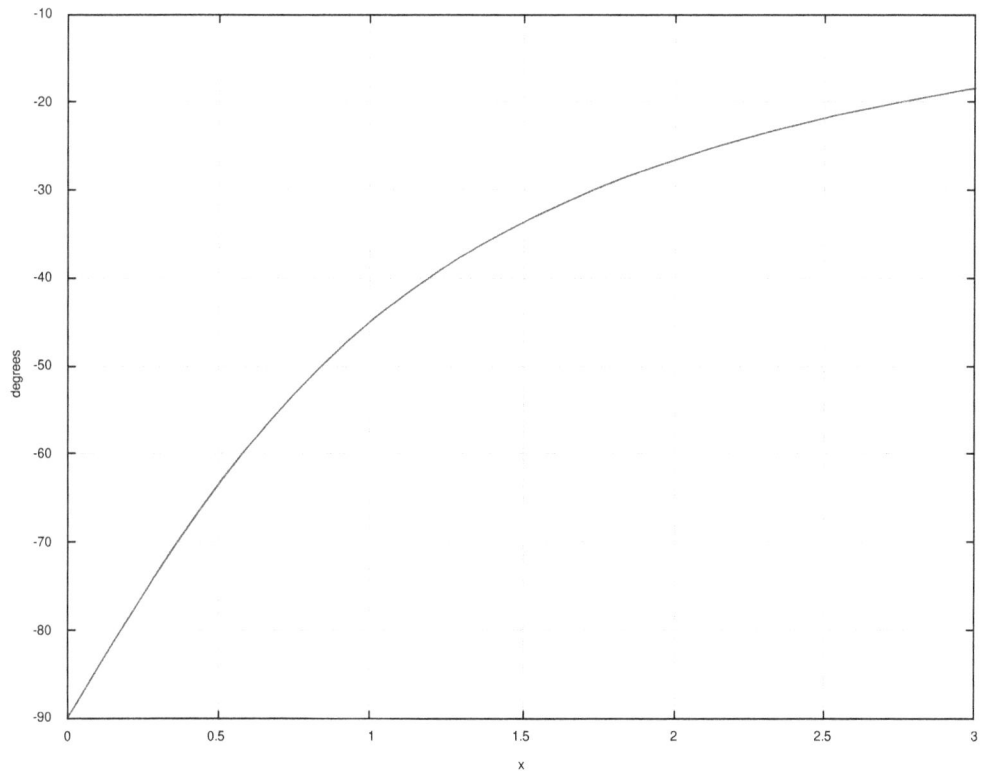

Figure 7: Phase response of first order high pass filter.

$$G(j\omega) = \frac{j\omega/\omega_0}{1 + j\omega/\omega_0} = \frac{jx}{1 + jx} = \frac{x}{\sqrt{1 + x^2}} e^{j(\frac{\pi}{2} - \arctan(x))} \tag{9}$$

$$g(x) = |G(j\omega)| = \frac{x}{\sqrt{1 + x^2}} \tag{10}$$

$$\theta(x) = \arg(G(j\omega)) = \frac{\pi}{2} - \arctan(x) \tag{11}$$

Figure 8 shows a first order high pass filter composed of one resistor and one inductor.

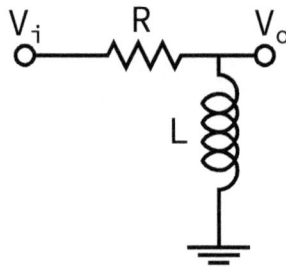

Figure 8: First order high pass filter using an inductor.

The frequency response is identical to the one for the RC high pass filter given above. The cutoff frequency in this case is $\omega_0 = R/L$.

The attenuation of these first order filters as a function of frequency is very gradual. For example at twice the cutoff frequency the output of the low pass filter is $V_0 = V_i/\sqrt{5} = 0.447V_i$. This is still almost half the input. Higher order filters that combine capacitors and inductors can produce much sharper attenuation. We'll start by looking at second order low pass, high pass, band pass and band stop filters. All these filters use one capacitor and one inductor in various arrangements.

SECOND ORDER FILTERS

An inductor is the electrical equivalent of a mass, and a capacitor is the electrical equivalent of a spring. Put the two of them together and you can get the phenomenon of resonance. This is where energy of a particular frequency is readily absorbed by the system while energy at other frequencies is rejected.

Combined with resistance, an inductor and capacitor combination is essentially a damped oscillator that will oscillate more strongly at some frequencies than at others. This frequency selective property is by definition a filter.

In this chapter we will show how to use one capacitor and one inductor to make what are called second order filters. Low pass, high pass, band pass and band stop filters will be discussed. They are called second order filters because the inductor and capacitor constitute two energy storage elements. The inductor stores energy in a magnetic field and the capacitor stores energy in an electric field, in the same way that a mass stores kinetic energy while a spring stores potential energy in an oscillating spring-mass system.

This ability to store energy in two different forms is essential for producing the phenomenon of resonance in a passive circuit. You can for example make a second order filter with just two capacitors or two inductors, but you won't get resonance in such a filter and the frequency selective property will be very poor. The capacitor and inductor combination allows for much sharper frequency responses that are sufficient for a lot of filtering applications. For even sharper responses multiple capacitors and inductors can be used. These will be discussed in following chapters.

Second Order Low Pass

Figure 9 shows a second order low pass filter with source resistance and no load resistance.

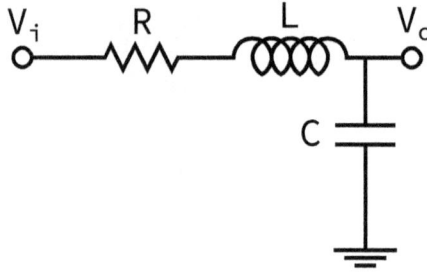

Figure 9: Second order low pass filter with source resistance.

The circuit is essentially a voltage divider with the output taken across the capacitance. At low frequencies the capacitor has high impedance and the inductor has low impedance. As the frequency increases the impedance of the inductor goes up and the impedance of the capacitor goes down. This is what produces the low pass property of the circuit. The transfer function for the filter is

$$G(s) = \frac{V_0}{V_i} = \frac{1}{LCs^2 + RCs + 1} \tag{12}$$

To get the frequency response from this function set $s = j\omega$ where $\omega = 2\pi f$ and f is the frequency in Hertz, then take the complex magnitude. Define the variable $x = \omega/\omega_0$ where ω_0 is the 3 dB cutoff frequency given by $\omega_0 = 1/\sqrt{LC}$. Define the parameter $a = R\sqrt{C/L}$. The frequency response can then be written in the following form.

$$g(a, x) = \frac{1}{\sqrt{(1 - x^2)^2 + a^2 x^2}} \tag{13}$$

A plot of the frequency response is shown in figure 10 for the values $a = 1, \sqrt{2}, 2$. Note that for $a = 1$ there is a peak in the response, for $a = \sqrt{2}$ the response stays more flat, and for $a = 2$ the response drops sharply.

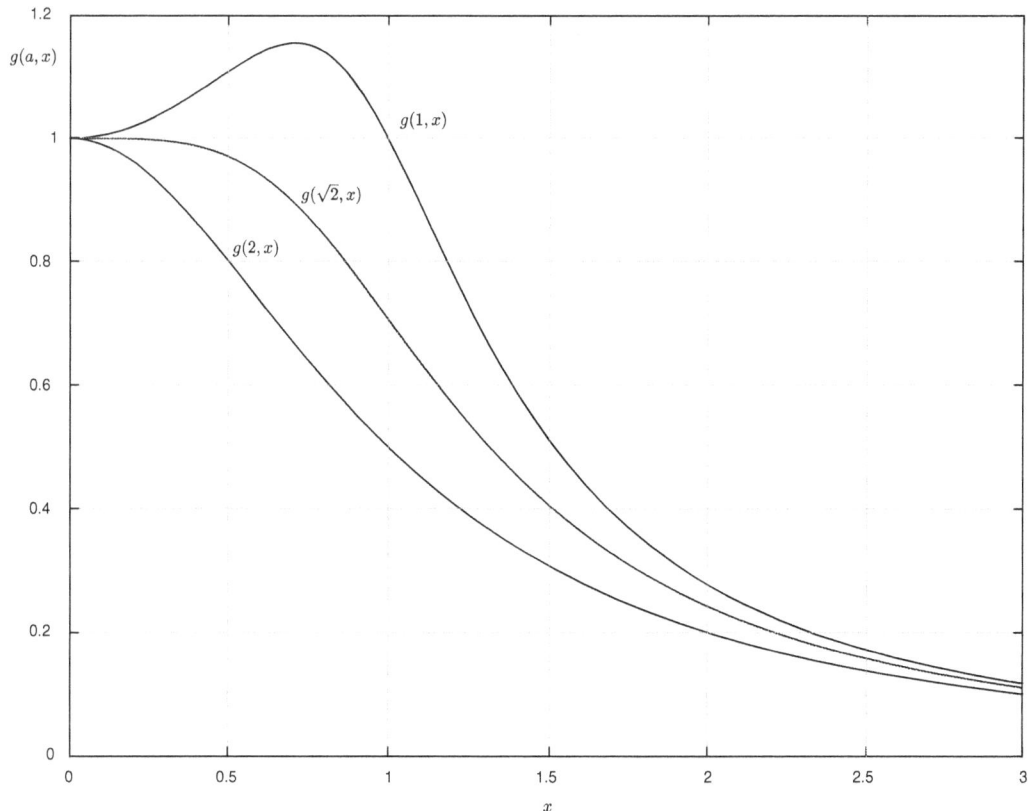

Figure 10: Second order low pass frequency response.

The reason for this behavior is clear if you look at the derivatives of $g(a, x)$ at $x = 0$. We have $g'(a, 0) = 0$ and $g''(a, 0) = 2 - a^2$. The second derivative is zero when $a = \sqrt{2}$ which causes the flat response. It is positive when $a < \sqrt{2}$ so we get a peak. It is negative when $a > \sqrt{2}$ so we get a sharp drop. The value $a = \sqrt{2}$ produces what is called a Butterworth filter also known as a maximally flat filter.

The phase difference between input and output is shown in figure 11 and is given by

$$\theta(a, x) = -\arctan\left(\frac{ax}{1 - x^2}\right) \tag{14}$$

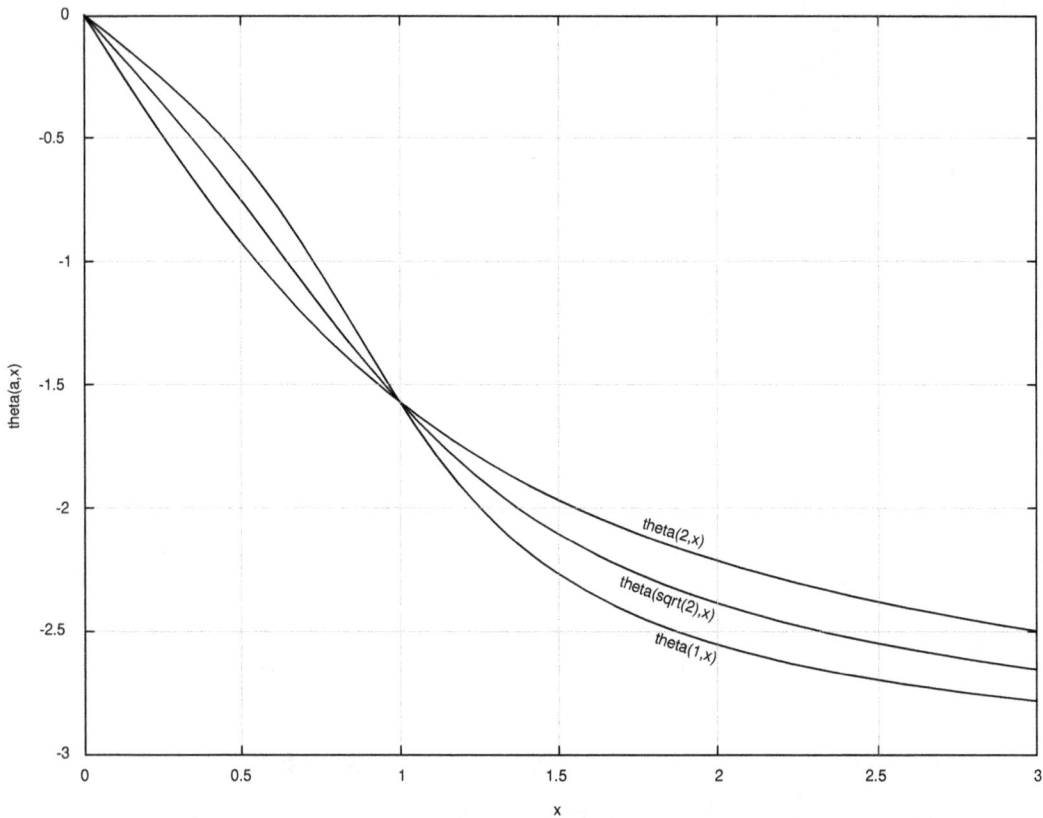

Figure 11: Second order low pass phase response.

Given the values of R, ω_0 and a, the design equations for this filter are

$$C = \frac{a}{R\omega_0} \tag{15}$$

$$L = \frac{R}{a\omega_0} \tag{16}$$

Figure 12 shows a second order low pass filter with load resistance and no source resistance.

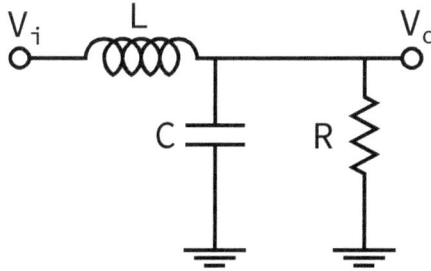

Figure 12: Second order low pass filter with load resistance.

The transfer function for this filter is

$$G(s) = \frac{V_0}{V_i} = \frac{1}{LCs^2 + \frac{L}{R}s + 1} \tag{17}$$

The frequency response is also given by equation 13 with the parameter a now defined as $\frac{1}{R}\sqrt{\frac{L}{C}}$. The frequency response curves are also the same and the maximally flat response also occurs when $a = \sqrt{2}$. Given R, ω_0 and a, the design equations for this filter are

$$C = \frac{1}{Ra\omega_0} \tag{18}$$

$$L = \frac{Ra}{\omega_0} \tag{19}$$

Second Order High Pass

Figure 13 shows a second order high pass filter with source resistance and no load resistance.

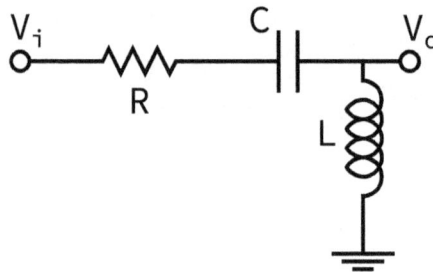

Figure 13: Second order high pass filter with source resistance.

We have a voltage divider with the output taken across the inductor. At low frequencies the capacitor has high impedance and the inductor has low impedance so the output is low. As the frequency increases the impedance of the inductor goes up and the impedance of the capacitor goes down so the output goes up. This produces the high pass property of the circuit. The transfer function for the filter is

$$G(s) = \frac{V_0}{V_i} = \frac{LCs^2}{LCs^2 + RCs + 1} \tag{20}$$

We get the frequency response the same way as for the low pass filter. Let $s = j\omega$ where $\omega = 2\pi f$ then take the complex magnitude. With the variable $x = \omega/\omega_0$, where ω_0 is the 3 dB cutoff frequency given by $\omega_0 = 1/\sqrt{LC}$, and the parameter $a = R\sqrt{C/L}$, the frequency response can be written in the following form.

$$g(a, x) = \frac{x^2}{\sqrt{(1 - x^2)^2 + a^2 x^2}} \tag{21}$$

A plot of the frequency response is shown in figure 14 for the values $a = 1, \sqrt{2}, 2$. Once again for $a = \sqrt{2}$ we get the flat Butterworth filter response.

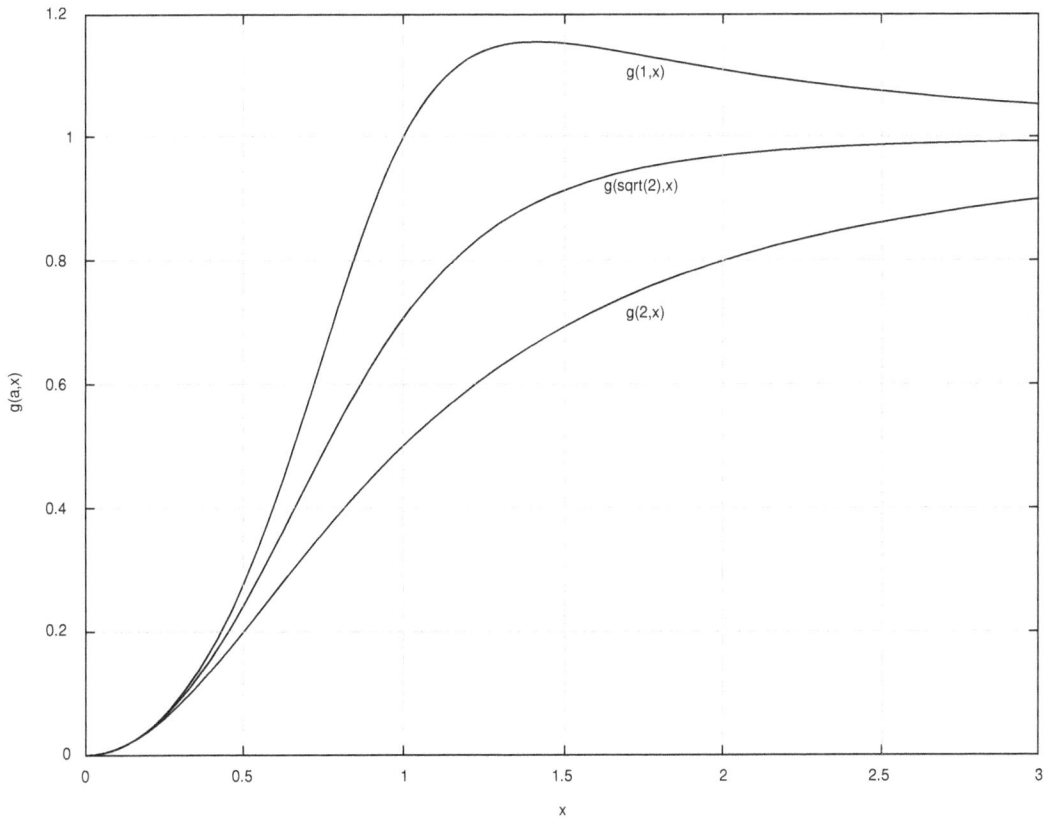

Figure 14: Second order high pass frequency response.

The phase difference between input and output is the same as for the low pass filter shown in figure 11.

The design equations are the same as for the low pass filter. The values for C and L are given by equations 15 and 16.

Figure 15 shows a second order high pass filter with load resistance and no source resistance.

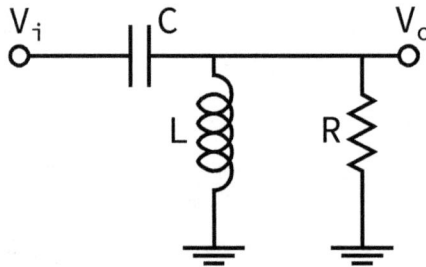

Figure 15: Second order high pass filter with load resistance.

The transfer function for this filter is

$$G(s) = \frac{V_0}{V_i} = \frac{LCs^2}{LCs^2 + \frac{L}{R}s + 1} \tag{22}$$

The frequency response is also given by equation 21 with the parameter a now defined as $\frac{1}{R}\sqrt{\frac{L}{C}}$. The design equations are the same as for the low pass filter. The values for C and L are given by equations 18 and 19.

Second Order Band Pass

Figure 16 shows a second order band pass filter with source resistance and no load resistance.

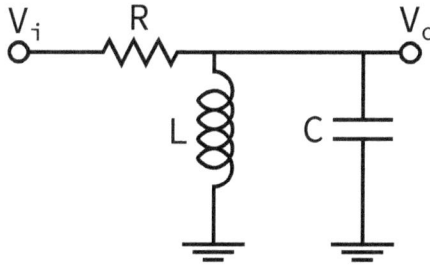

Figure 16: Second order band pass filter with source resistance and no load resistance.

Let Z be the impedance of the parallel LC combination. The resistor and Z then form a voltage divider. At low frequencies the inductor looks like a short and the output goes to zero. At high frequencies the capacitor looks like a short and the output goes to zero. In between, at the frequency $\omega_0 = \frac{1}{\sqrt{LC}}$, the value of Z goes to infinity and $V_o = V_i$. We therefore have a band pass filter with a response that peaks at ω_0. The transfer function for the filter is

$$G(s) = \frac{V_0}{V_i} = \frac{\frac{L}{R}s}{LCs^2 + \frac{L}{R}s + 1} \tag{23}$$

We can write the frequency response for the filter in the following form.

$$g(a,x) = \frac{ax}{\sqrt{(1-x^2)^2 + a^2 x^2}} \tag{24}$$

where $x = \frac{\omega}{\omega_0}$, $\omega_0 = \frac{1}{\sqrt{LC}}$ and $a = \frac{1}{R}\sqrt{\frac{L}{C}}$. A plot of the frequency response for the values $a = 1, 0.5, 0.1$ is shown in figure 17.

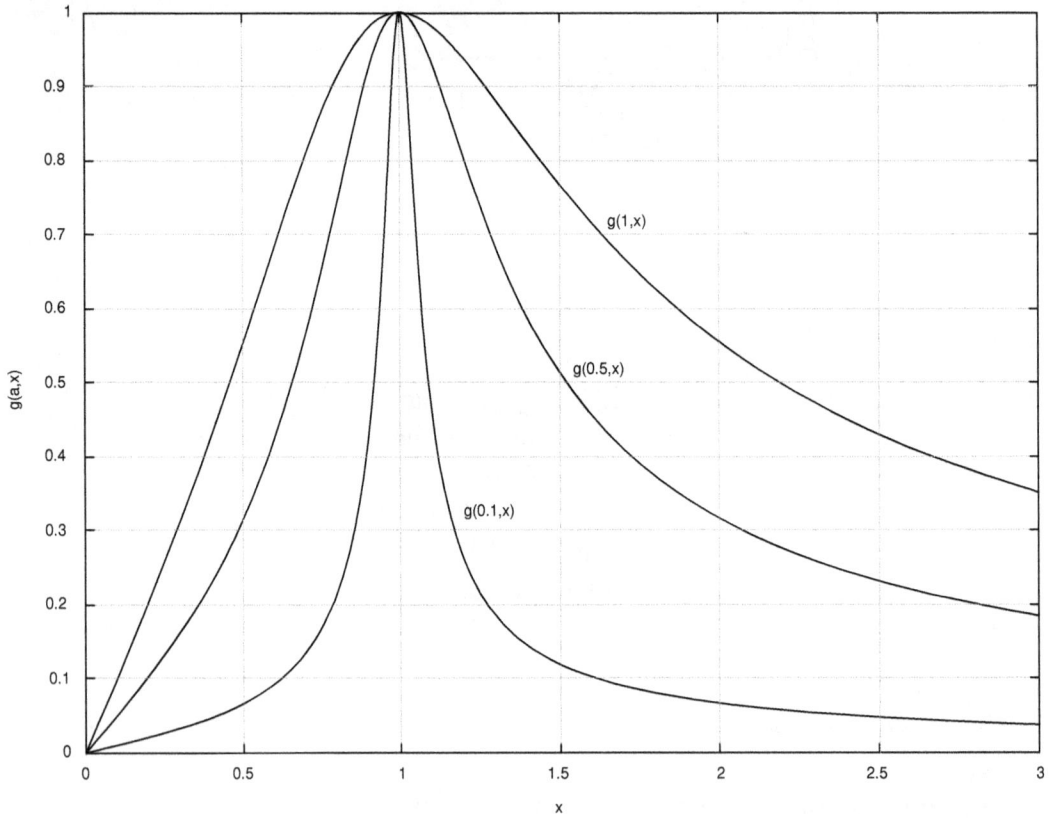

Figure 17: Second order band pass frequency response.

The frequency response always peaks at $x = 1$ or $\omega = \omega_0$ and $g(a, 1) = 1$ for all values of a. Note that as a decreases, the bandwidth of the response narrows. The exact value of the bandwidth is found from the values of x where $g(a, x) = 1/\sqrt{2}$. Those values are

$$x_1 = \sqrt{1 + \frac{a^2}{4}} - \frac{a}{2} \qquad\qquad x_2 = \sqrt{1 + \frac{a^2}{4}} + \frac{a}{2} \tag{25}$$

The normalized (unitless) bandwidth is then

$$\frac{\text{BW}}{\omega_0} = x_2 - x_1 = a \tag{26}$$

or in terms of frequency we have $\text{BW} = a\omega_0$.

The phase difference between input and output is shown in figure 18 and is given by

$$\theta(a, x) = \frac{\pi}{2} - \arctan\left(\frac{ax}{1 - x^2}\right) \tag{27}$$

To design this filter for a bandwidth BW and center frequency ω_0 let $a = \text{BW}/\omega_0$ and use the following values for L and C

$$C = \frac{1}{Ra\omega_0} \tag{28}$$

$$L = \frac{Ra}{\omega_0} \tag{29}$$

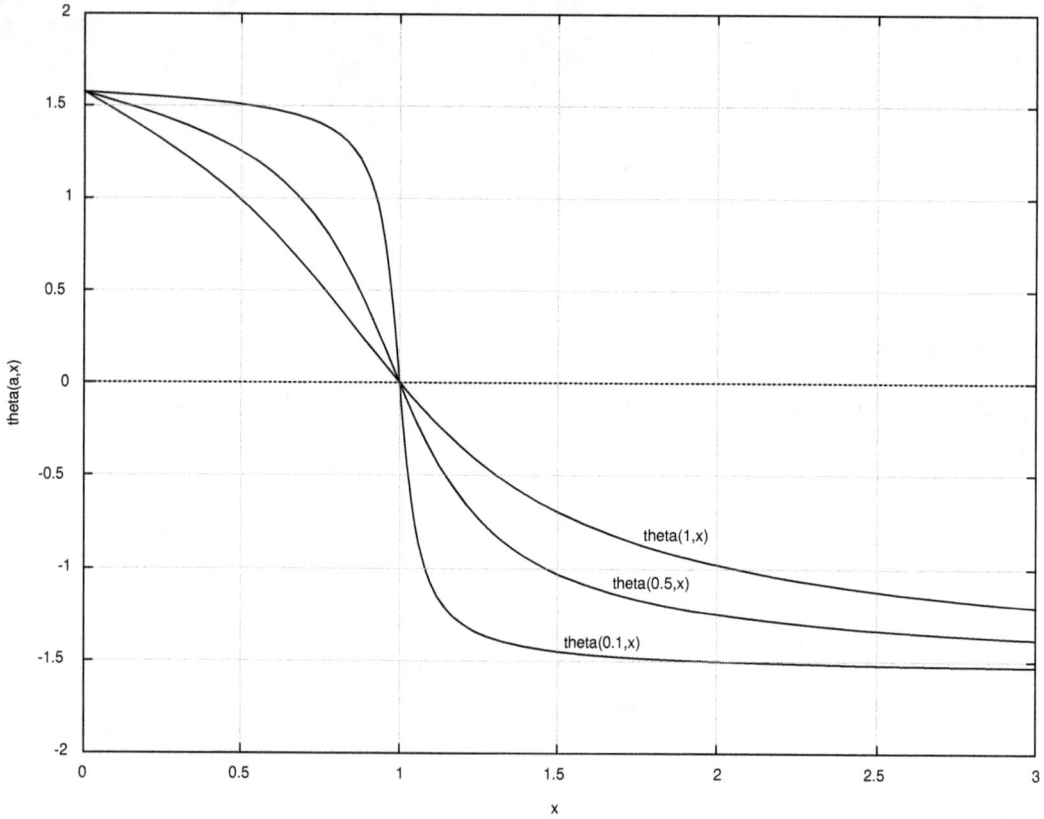

Figure 18: Second order band pass phase response.

Second Order Band Stop

Figure 19 shows a second order band stop filter with source resistance and no load resistance.

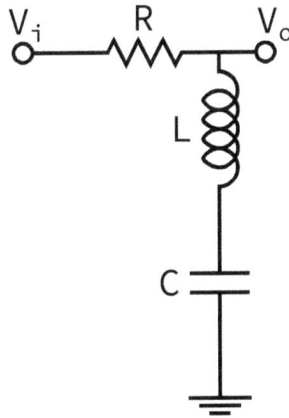

Figure 19: Second order band stop filter with source resistance.

The resistor and series LC combination form a voltage divider. At low frequencies the capacitor looks like an open circuit while at high frequencies the inductor looks like an open circuit. At high and low frequencies then we have $V_o \approx V_i$. At the frequency $\omega_0 = 1/\sqrt{LC}$ the impedance of the series LC combination goes to zero and $V_o = 0$. We therefore have a band stop filter with a zero at ω_0. The transfer function for the filter is

$$G(s) = \frac{V_0}{V_i} = \frac{LCs^2 + 1}{LCs^2 + RCs + 1} \tag{30}$$

We can write the frequency response for the filter in the following form.

$$g(a, x) = \frac{|1 - x^2|}{\sqrt{(1 - x^2)^2 + a^2 x^2}} \tag{31}$$

where $x = \frac{\omega}{\omega_0}$, $\omega_0 = 1/\sqrt{LC}$ and $a = R\sqrt{\frac{C}{L}}$. A plot of the frequency response for the values $a = 1, 0.5, 0.1$ is shown in figure 20.

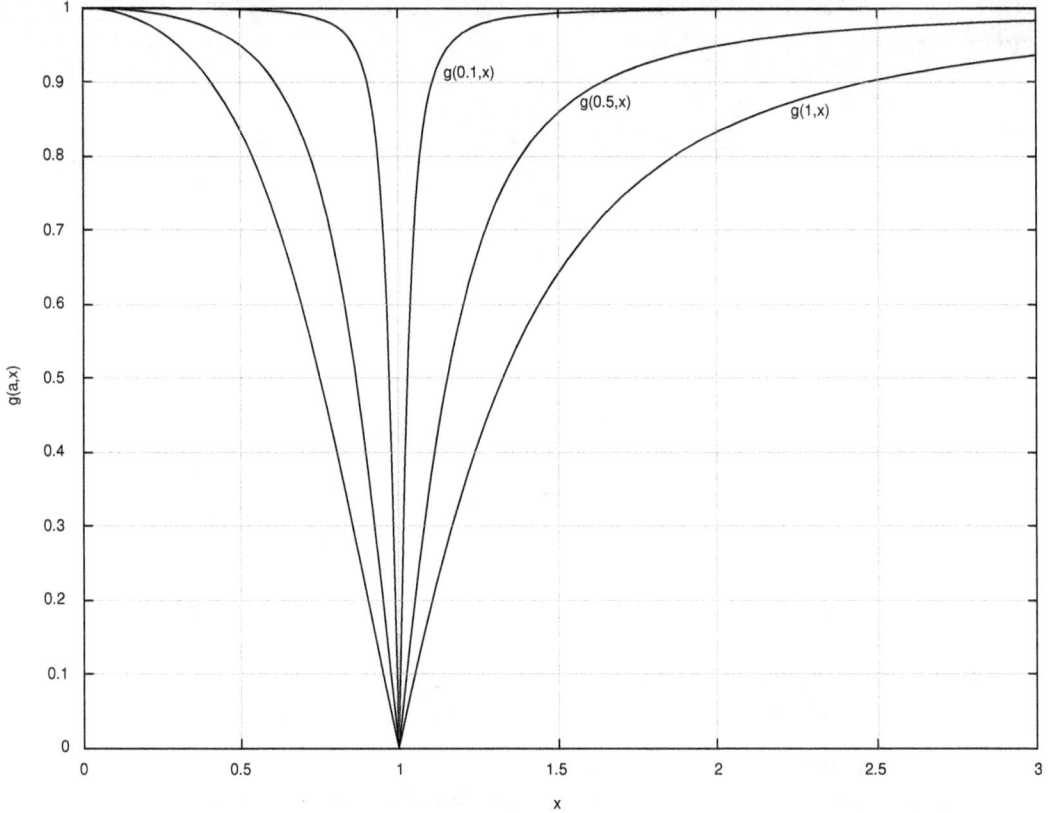

Figure 20: Second order band stop frequency response.

The frequency response always goes to zero at $x = 1$ or $\omega = \omega_0$ and $g(a, 1) = 1$ for all values of a. Note that as a decreases the bandwidth of the response narrows. The exact value of the bandwidth is found from the values of x where $g(a, x) = 1/\sqrt{2}$. Those values are

$$x_1 = \sqrt{1 + \frac{a^2}{4}} - \frac{a}{2} \qquad\qquad x_2 = \sqrt{1 + \frac{a^2}{4}} + \frac{a}{2} \qquad\qquad (32)$$

The normalized (unitless) bandwidth is then

$$\frac{\text{BW}}{\omega_0} = x_2 - x_1 = a \tag{33}$$

or in terms of frequency we have $\text{BW} = a\omega_0$.

The phase difference between input and output is shown in figure 21 and is given by

$$\theta(a, x) = -\arctan\left(\frac{ax}{1 - x^2}\right) \tag{34}$$

To design this filter for a bandwidth BW and stop frequency ω_0 let $a = \text{BW}/\omega_0$ and use the following values for L and C

$$C = \frac{a}{R\omega_0} \tag{35}$$

$$L = \frac{R}{a\omega_0} \tag{36}$$

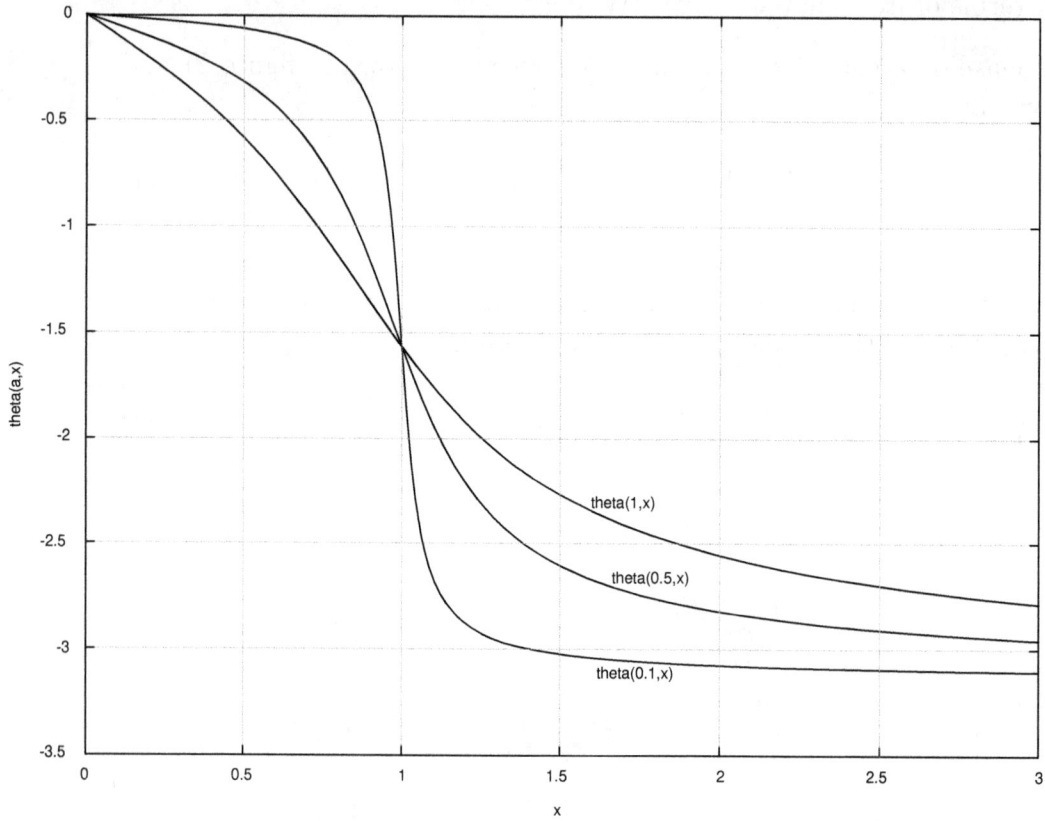

Figure 21: Second order band stop phase response.

BUTTERWORTH

The Butterworth filter is named after Stephen Butterworth (1885–1958) a British physicist who introduced his namesake filter to the world in 1930 in his paper "On the Theory of Filter Amplifiers" (Experimental Wireless and the Wireless Engineer, Vol 7, Oct 1930, pg 536-541).

A Butterworth filter is also called a maximally flat filter because its frequency response has no ripples in the pass or stop bands. The high pass, band pass, and band stop Butterworth filters can all be derived from the low pass filter so we will start by taking a close look at the low pass filter.

The component values for all the filters in this book are ultimately derived from the poles of the low pass Butterworth functions. So if you're interested in where these values come from you may want to read through the following discussion. But the material is not needed if you simply want to design a filter. In that case find the filter you want in the following pages and scale the component values to the desired frequency using the given recipe.

The magnitude of the frequency response for the nth order low pass Butterworth filter is usually written as follows

$$|G_n(j\omega)| = \frac{1}{\sqrt{1 + \epsilon^2 \omega^{2n}}} \tag{37}$$

Plots of the frequency response for $\epsilon = 1$ and $n = 2, 4, 6, 8, 10$ are shown in figure 22.

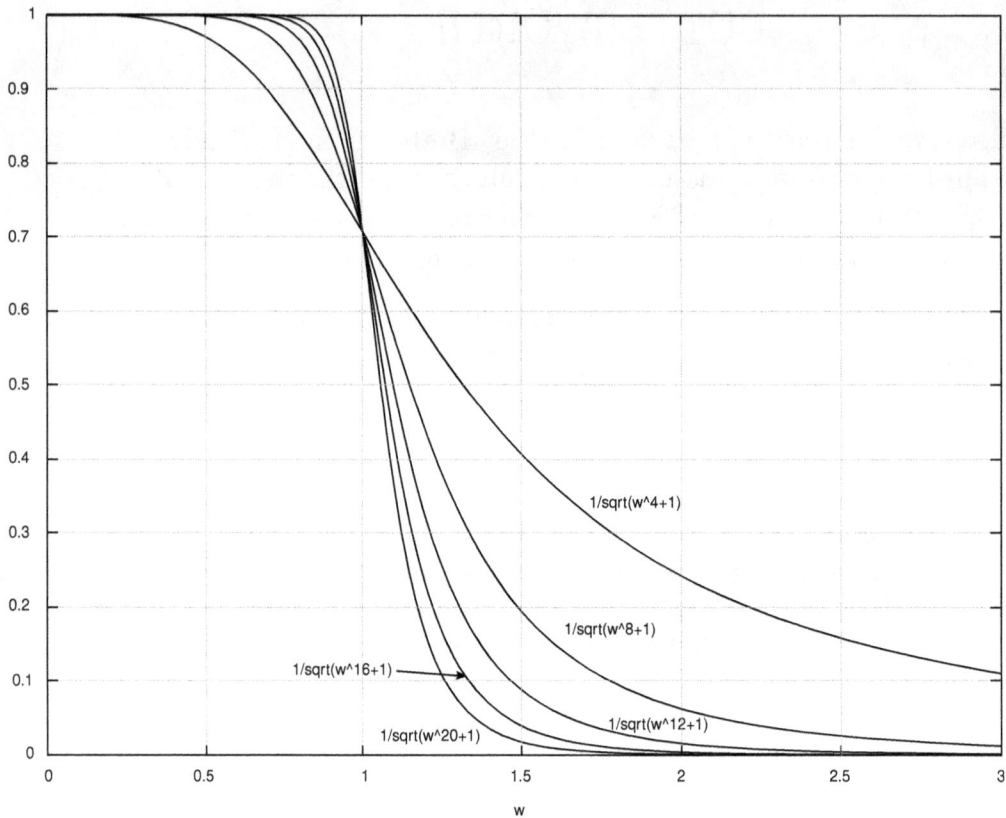

Figure 22: Butterworth frequency response for $n = 2, 4, 6, 8, 10$.

This is a normalized response where the cutoff frequency (end of the pass band) is $\omega = 1$. For $\epsilon = 1$ the magnitude of the response at $\omega = 1$ is $1/\sqrt{2}$ for all values of n. The same graphs hold for any cutoff frequency ω_0 by simply multiplying the frequency axis by ω_0.

Plots of the phase response for $\epsilon = 1$ and $n = 2, 4, 6, 8, 10$ are shown in figure 23.

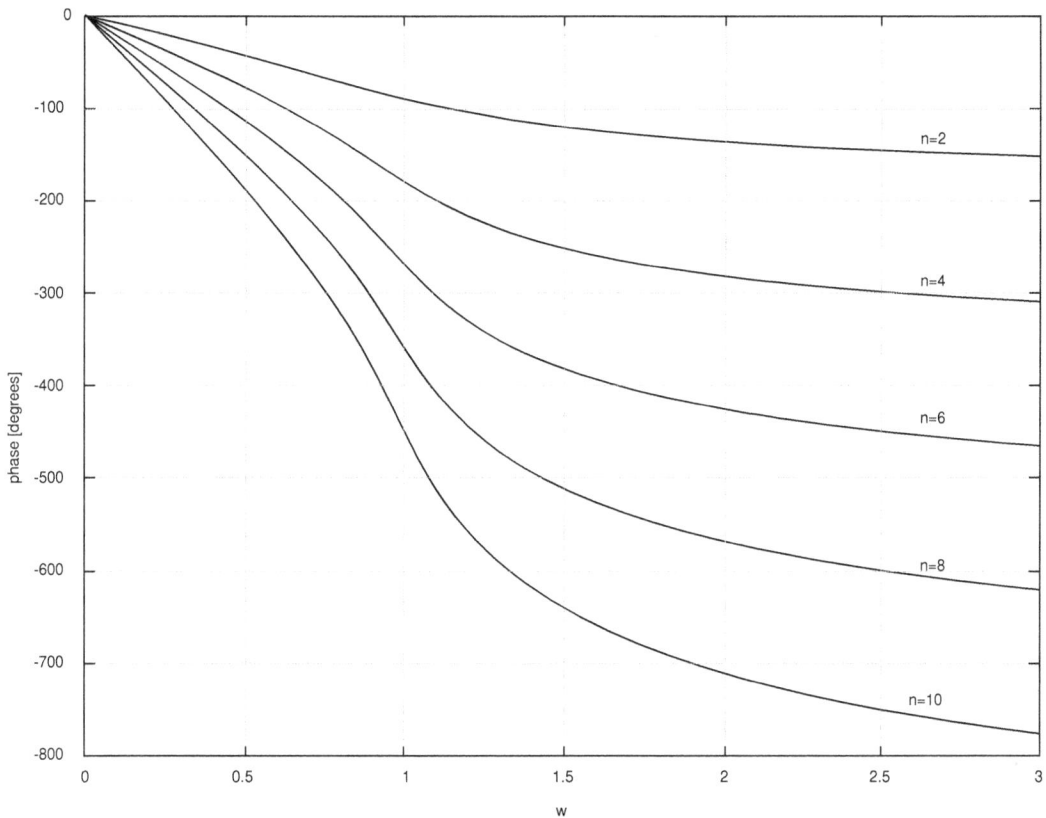

Figure 23: Butterworth phase response for $n = 2, 4, 6, 8, 10$.

The parameter ϵ can be used to set the magnitude of the response at $\omega = 1$ to something other than $1/\sqrt{2}$ since at $\omega = 1$ the response is

$$|G_n(j)| = \frac{1}{\sqrt{1 + \epsilon^2}} \tag{38}$$

In our view the parameter ϵ is superfluous and we will use $\epsilon = 1$ in everything that follows. One can then simply choose the cutoff frequency to be the frequency at which the magnitude of the response should equal $1/\sqrt{2}$.

To design any filter the transfer function for the filter must be specified in terms of the Laplace transform variable $s = j\omega$. In the case of a Butterworth filter the transfer function is used to produce a continued fraction expansion of either the input or output impedance or admittance functions. The component values can then be read off from the expansion.

If $G(s)$ is the transfer function then the square of the frequency response is related to it as follows:

$$|G(j\omega)|^2 = G(j\omega)G(-j\omega) = G(s)G(-s) \tag{39}$$

So if we substitute $\omega^2 = -s^2$ into

$$|G_n(j\omega)|^2 = \frac{1}{1 + \omega^{2n}} \tag{40}$$

we get an expression for the product $G_n(s)G_n(-s)$. The denominator of this product is $1 + (-s^2)^n$. To find the poles of the filter we set this equal to zero and solve for s. The solutions in the left half plane (negative real part) will be the poles of $G_n(s)$.

The nth order Butterworth system function can then be written as

$$G_n(s) = \prod_{k=0}^{n-1} \frac{1}{(s - p_k)} \tag{41}$$

where p_k are the poles given by the following equation

$$p_k = -\sin\frac{\pi(2k+1)}{2n} + j\cos\frac{\pi(2k+1)}{2n}$$
$$k = 0, 1, 2, \ldots, n-1 \tag{42}$$

The poles always come in complex conjugate pairs, so the system function can also be written in the following form

$$G_n(s) = \frac{1}{D_n(s)} \quad n = 1, 2, 3 \ldots \tag{43}$$

$$D_n(s) = \begin{cases} (s+1) \displaystyle\prod_{k=0}^{(n-3)/2} (s^2 + 2r_k s + 1) & n = 1, 3, 5, \ldots \\ \displaystyle\prod_{k=0}^{(n-2)/2} (s^2 + 2r_k s + 1) & n = 2, 4, 6, \ldots \end{cases}$$

$$r_k = \sin\left(\frac{\pi(2k+1)}{2n}\right) \tag{44}$$

The system function for orders 1 through 6 are shown below.

$$G_1(s) = \frac{1}{s+1} \tag{45}$$

$$G_2(s) = \frac{1}{s^2 + \sqrt{2}s + 1} \tag{46}$$

$$G_3(s) = \frac{1}{s^3 + 2s^2 + 2s + 1} \tag{47}$$

$$G_4(s) = \frac{1}{s^4 + \sqrt{4 + \sqrt{8}}s^3 + (2 + \sqrt{2})s^2 + \sqrt{4 + \sqrt{8}}s + 1} \tag{48}$$

$$G_5(s) = \frac{1}{(s+1)(s^4 + \sqrt{5}s^3 + 3s^2 + \sqrt{5}s + 1)} \tag{49}$$

$$G_6(s) = \frac{1}{(s^2 + \sqrt{2}s + 1)(s^4 + \sqrt{6}s^3 + 3s^2 + \sqrt{6}s + 1)} \tag{50}$$

We will now outline briefly how these functions can be used to determine filter component values.

The filters are implemented as LC ladder networks with all the series elements being inductors and all shunt elements being capacitors. In the case of finite source resistance and infinite load resistance, if the first element is a series inductor then the impedance of the ladder can be written down by inspection as a continued fraction. For example the impedance of the fourth order LC ladder in figure 33 is

$$Z = L_1 s + \cfrac{1}{C_1 s + \cfrac{1}{L_2 s + \cfrac{1}{C_2 s}}} \tag{51}$$

If the first element is a shunt capacitor then the admittance can be written down as a continued fraction. For example the admittance of the third order LC ladder in figure 29 is

$$Y = C_1 s + \cfrac{1}{Ls + \cfrac{1}{C_2 s}} \tag{52}$$

The impedance and admittance for higher order filters follow the same pattern.

Now we can get the component values in equations 51 and 52 using the $G_4(s)$ and $G_3(s)$ Butterworth functions given above. For the fourth order filter we have

$$Z = \frac{s^4 + +(2 + \sqrt{2})s^2 + 1}{\sqrt{4 + \sqrt{8}}s^3 + \sqrt{4 + \sqrt{8}}s} \tag{53}$$

i.e. the impedance of the fourth order filter is equal to the ratio of a polynomial of the even terms to a polynomial of the odd terms in the denominator of $G_4(s)$. Expanding equation 53 as a continued fraction gives you the component values.

Likewise for the third order filter we have

$$Y = \frac{s^3 + 2s}{2s^2 + 1} \tag{54}$$

Expanding this in a continued fraction gives you the component values for the filter. The expansion is

$$Y = \frac{1}{2}s + \cfrac{1}{\frac{4}{3}s + \cfrac{1}{\frac{3}{2}s}} \tag{55}$$

Comparing this with equation 52 we see that $C_1 = 1/2$, $L = 4/3$, $C_2 = 3/2$.

The same basic procedure applies in the case of finite load resistance and zero source resistance. The only difference is that the impedance or admittance of the ladder is calculated looking into the output side of the ladder with the source shorted. The same continued fraction expansion of the Butterworth function is used to get the component values. The component values are therefore the same but assigned in opposite order to the values for the filter with source resistance.

For filters with equal source and load resistance the component values can be calculated directly. If e_k is the value of the kth element in the ladder then

$$e_k = 2\sin\frac{(2k-1)\pi}{2n} \tag{56}$$

where $k = 1, 2, 3, \ldots, n$.

In what follows we provide scalable L and C values for high and low pass filter orders 1 through 10. These values were calculated using the procedures outlined above. Scalable means that if you want to set the 3 dB frequency to ω_0, then you divide each value by $\omega_0 = 2\pi f_0$. Each filter has either a source or load resistor whose default value is 1Ω. If you want to change the resistance value to R, then multiply each inductance by R, and divide each capacitance by R.

Low Pass Butterworth

Circuit diagrams for the second through tenth order low pass filters are shown on the following pages. The given component values will produce a low pass filter with $R = 1$ and $\omega_0 = 1$.

The component values in these circuits can easily be converted to other values of R and ω_0 by scaling them as follows

$$L \to \frac{LR}{\omega_0} \qquad\qquad C \to \frac{C}{R\omega_0} \qquad\qquad (57)$$

For example if we want a fourth order low pass filter with cutoff at 20 kHz and a source resistance of 50 Ω then using the component values in equation 67 we get, for $R = 50\Omega$, and $\omega_0 = 2\pi \cdot 20000$,

$$
\begin{aligned}
L_1 &\to L_1 R/\omega_0 & &\approx 152\mu\text{H} \\
C_1 &\to C_1/R\omega_0 & &\approx 172\text{nF} \\
C_2 &\to C_2/R\omega_0 & &\approx 244\text{nF} \\
L_2 &\to L_2 R/\omega_0 & &\approx 627.5\mu\text{H}
\end{aligned}
$$

$$(58)$$

The spice code and the frequency response simulation for the filter are shown below.

```
.title low pass 4th order with source resistance
V1 in 0 dc 0 ac 1
R1 in 1 50
L1 1 2 152u
C1 2 0 172n
L2 2 out 627.5u
```

```
C2 out 0 244n
.control
ac dec 1000 10000 50000
gnuplot newplot vm(out)
.endc
.end
```

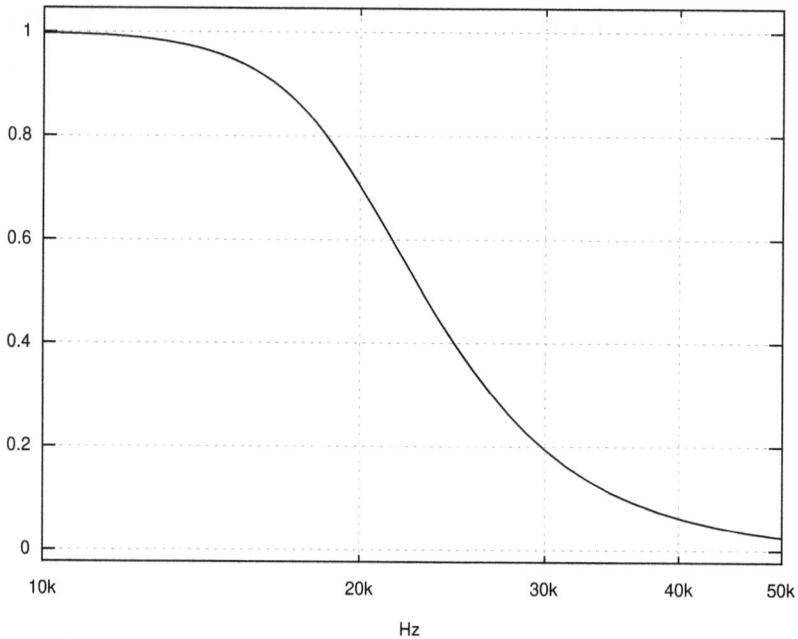

Figure 24: Spice generated frequency response for fourth order low pass filter with source resistance $R = 50\Omega$ and $L_1 = 152\mu H$, $C_1 = 172nF$, $C_2 = 244nF$ and $L_2 = 627.5\mu H$.

Second Order Low Pass Butterworth

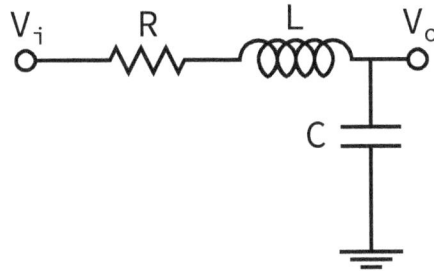

Figure 25: Second order low pass Butterworth with source impedance.

$$C = \sqrt{2} \qquad\qquad L = 1/\sqrt{2} \qquad\qquad (59)$$

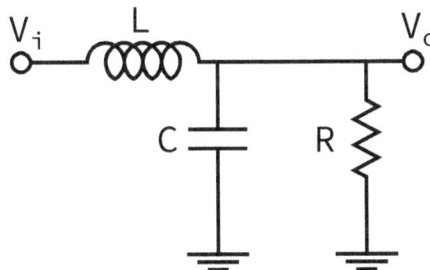

Figure 26: Second order low pass Butterworth with load impedance.

$$C = 1/\sqrt{2} \qquad\qquad L = \sqrt{2} \qquad\qquad (60)$$

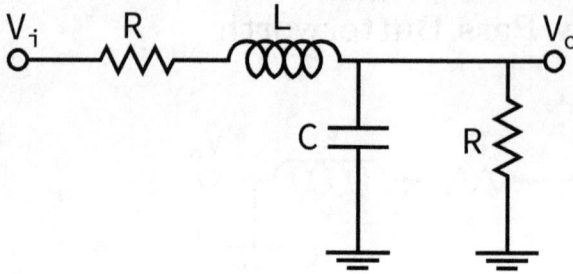

Figure 27: Second order low pass Butterworth with source and load impedance, variation 1.

$$C = \sqrt{2} \qquad\qquad L = \sqrt{2} \qquad\qquad (61)$$

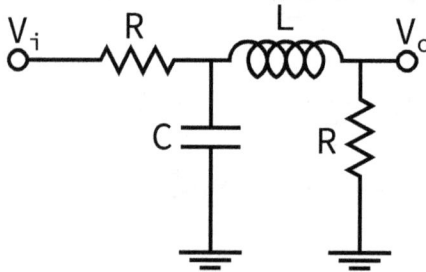

Figure 28: Second order low pass Butterworth with source and load impedance, variation 2.

$$C = \sqrt{2} \qquad\qquad L = \sqrt{2} \qquad\qquad (62)$$

Third Order Low Pass Butterworth

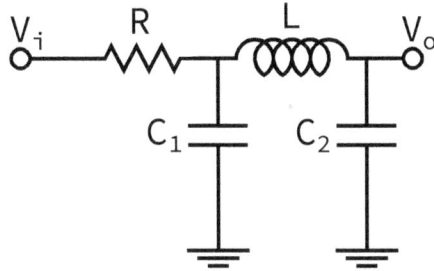

Figure 29: Third order low pass Butterworth with source impedance.

$$C_1 = 1/2 \qquad L = 4/3 \qquad C_2 = 3/2 \qquad (63)$$

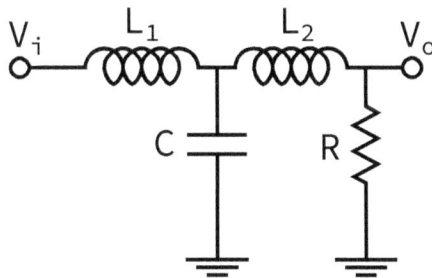

Figure 30: Third order low pass Butterworth with load impedance.

$$L_1 = 3/2 \qquad C = 4/3 \qquad L_2 = 1/2 \qquad (64)$$

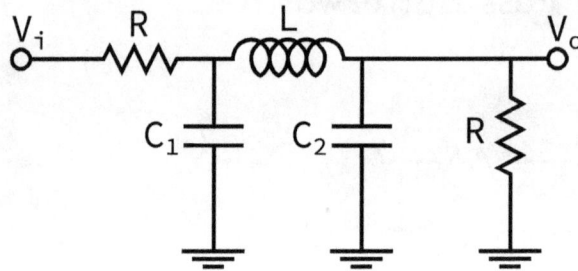

Figure 31: Third order low pass Butterworth with source and load impedance, variation 1.

$$C_1 = 1 \qquad L = 2 \qquad C_2 = 1 \tag{65}$$

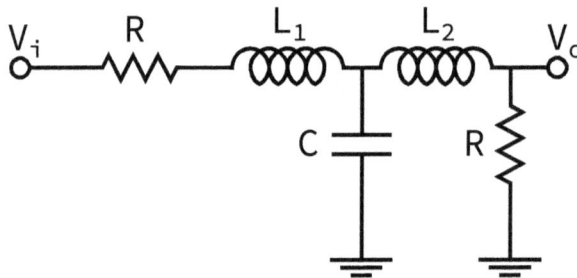

Figure 32: Third order low pass Butterworth with source and load impedance, variation 2.

$$L_1 = 1 \qquad C = 2 \qquad L_2 = 1 \tag{66}$$

Fourth Order Low Pass Butterworth

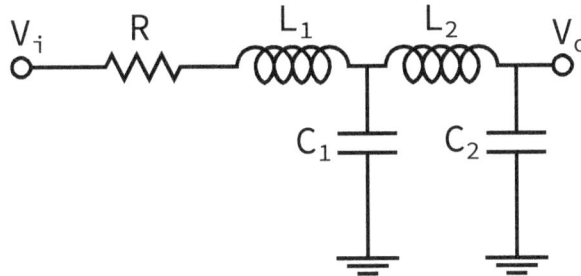

Figure 33: Fourth order low pass Butterworth with source impedance.

$$L_1 = 0.3826834324 \qquad C_1 = 1.0823922003 \qquad (67)$$
$$L_2 = 1.5771610149 \qquad C_2 = 1.5307337295$$

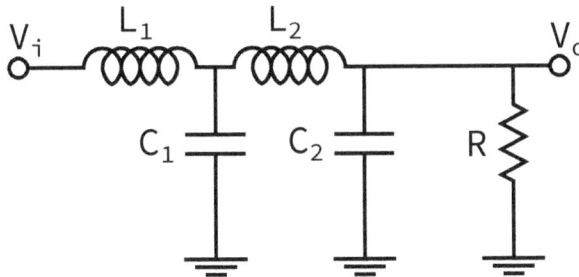

Figure 34: Fourth order low pass Butterworth with load impedance.

$$L_1 = 1.5307337295 \qquad C_1 = 1.5771610149 \qquad (68)$$
$$L_2 = 1.0823922003 \qquad C_2 = 0.3826834324$$

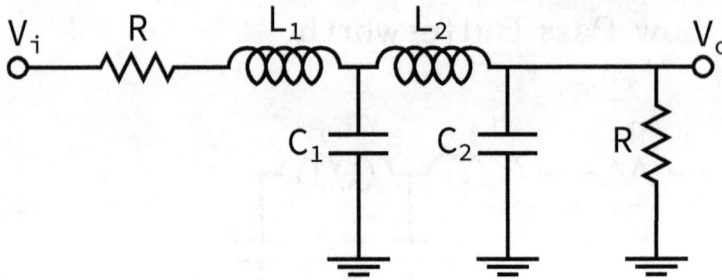

Figure 35: Fourth order low pass Butterworth with source and load impedance, variation 1.

$$L_1 = 0.7653668647 \qquad C_1 = 1.8477590650 \tag{69}$$
$$L_2 = 1.8477590650 \qquad C_2 = 0.7653668647$$

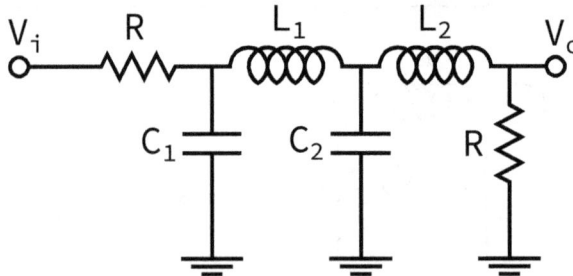

Figure 36: Fourth order low pass Butterworth with source and load impedance, variation 2.

$$C_1 = 0.7653668647 \qquad L_1 = 1.8477590650 \tag{70}$$
$$C_2 = 1.8477590650 \qquad L_2 = 0.7653668647$$

Fifth Order Low Pass Butterworth

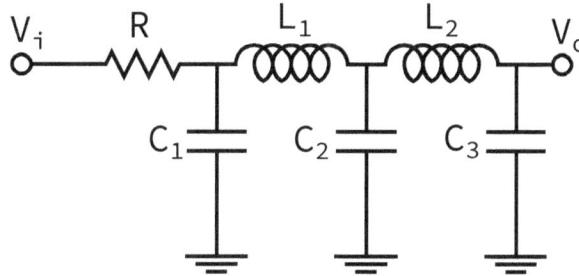

Figure 37: Fifth order low pass Butterworth with source impedance.

$$C_1 = 0.3090169944 \qquad\qquad L_1 = 0.8944271910 \qquad\qquad (71)$$
$$C_2 = 1.3819660113 \qquad\qquad L_2 = 1.6944271910$$
$$C_3 = 1.5450849719$$

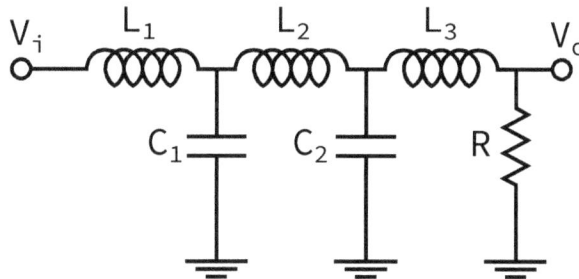

Figure 38: Fifth order low pass Butterworth with load impedance.

$$L_1 = 1.5450849719 \qquad\qquad C_1 = 1.6944271910 \qquad\qquad (72)$$
$$L_2 = 1.3819660113 \qquad\qquad C_2 = 0.8944271910$$
$$L_3 = 0.3090169944$$

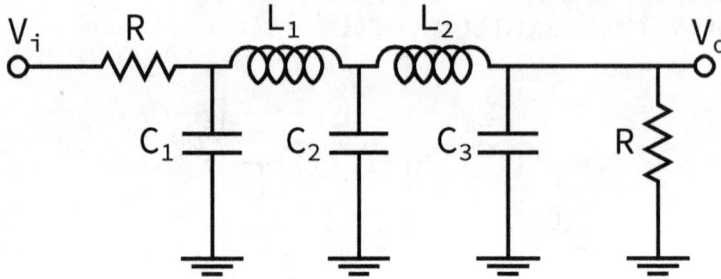

Figure 39: Fifth order low pass Butterworth with source and load impedance, variation 1.

$$C_1 = 0.6180339887 \qquad\qquad L_1 = 1.6180339887 \qquad\qquad (73)$$
$$C_2 = 2.0 \qquad\qquad\qquad L_2 = 1.6180339887$$
$$C_3 = 0.6180339887$$

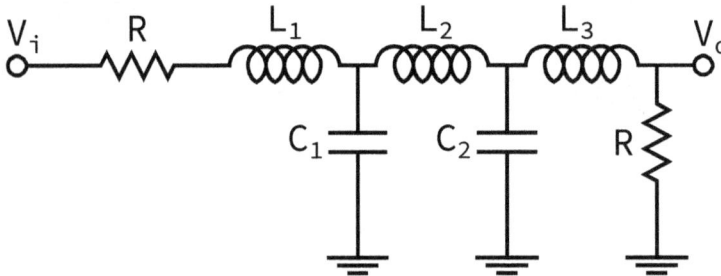

Figure 40: Fifth order low pass Butterworth with source and load impedance, variation 2.

$$L_1 = 0.6180339887 \qquad\qquad C_1 = 1.6180339887 \qquad\qquad (74)$$
$$L_2 = 2.0 \qquad\qquad\qquad C_2 = 1.6180339887$$
$$L_3 = 0.6180339887$$

Sixth Order Low Pass Butterworth

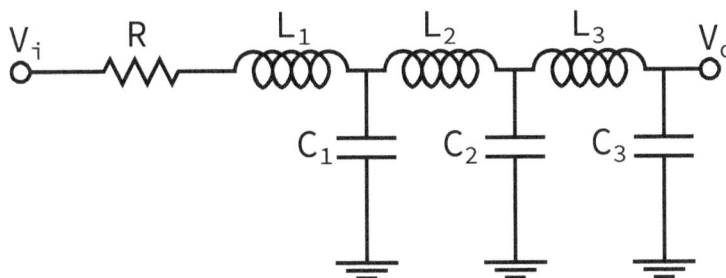

Figure 41: Sixth order low pass Butterworth with source impedance.

$$L_1 = 0.2588190451 \qquad C_1 = 0.7578747639 \qquad (75)$$
$$L_2 = 1.2016280867 \qquad C_2 = 1.5529142706$$
$$L_3 = 1.7593056225 \qquad C_3 = 1.5529142706$$

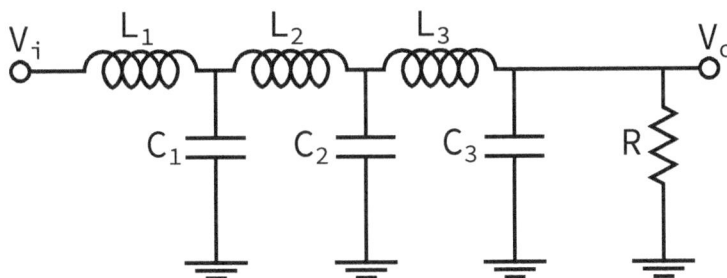

Figure 42: Sixth order low pass Butterworth with load impedance.

$$L_1 = 1.5529142706 \qquad C_1 = 1.7593056225 \qquad (76)$$
$$L_2 = 1.5529142706 \qquad C_2 = 1.2016280867$$
$$L_3 = 0.7578747639 \qquad C_3 = 0.2588190451$$

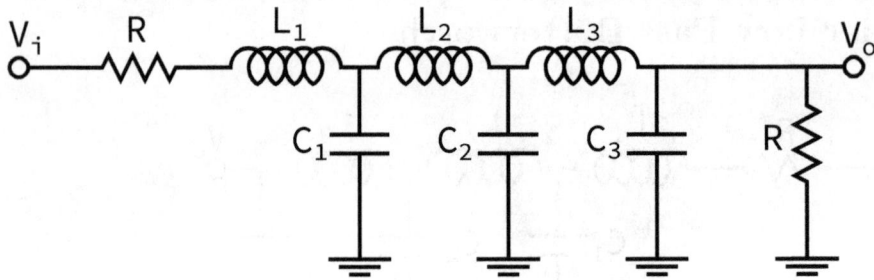

Figure 43: Sixth order low pass Butterworth with source and load impedance, variation 1.

$$L_1 = 0.5176380902 \qquad C_1 = 1.4142135624 \qquad (77)$$
$$L_2 = 1.9318516526 \qquad C_2 = 1.9318516526$$
$$L_3 = 1.4142135624 \qquad C_3 = 0.5176380902$$

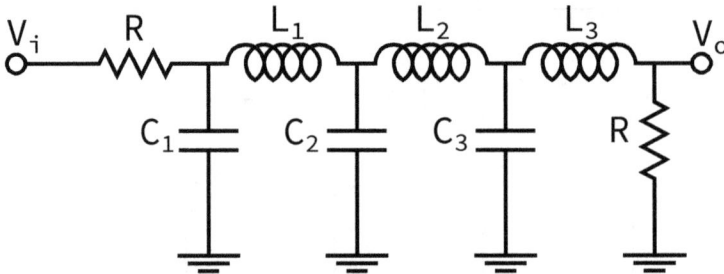

Figure 44: Sixth order low pass Butterworth with source and load impedance, variation 2.

$$C_1 = 0.5176380902 \qquad L_1 = 1.4142135624 \qquad (78)$$
$$C_2 = 1.9318516526 \qquad L_2 = 1.9318516526$$
$$C_3 = 1.4142135624 \qquad L_3 = 0.5176380902$$

Seventh Order Low Pass Butterworth

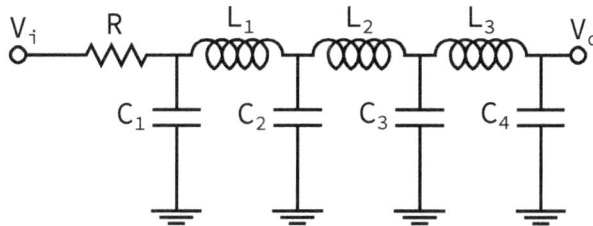

Figure 45: Seventh order low pass Butterworth with source impedance.

$$
\begin{aligned}
C_1 &= 0.2225209340 & L_1 &= 0.6559705552 \\
C_2 &= 1.0549581321 & L_2 &= 1.3971667817 \\
C_3 &= 1.6588336037 & L_3 &= 1.7988276981 \\
C_4 &= 1.5576465377 &
\end{aligned}
\tag{79}
$$

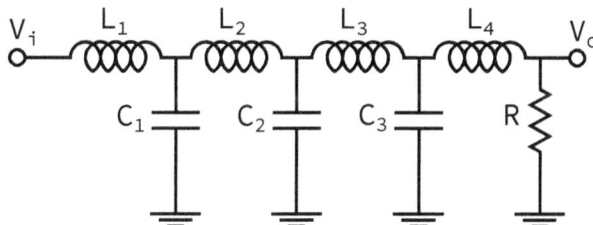

Figure 46: Seventh order low pass Butterworth with load impedance.

$$
\begin{aligned}
L_1 &= 1.5576465377 & C_1 &= 1.7988276981 \\
L_2 &= 1.6588336037 & C_2 &= 1.3971667817 \\
L_3 &= 1.0549581321 & C_3 &= 0.655970555 \\
L_4 &= 0.2225209340 &
\end{aligned}
\tag{80}
$$

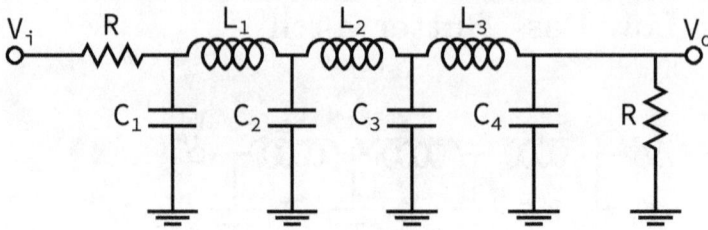

Figure 47: Seventh order low pass Butterworth with source and load impedance, variation 1.

$$C_1 = 0.4450418679 \qquad L_1 = 1.2469796037 \qquad (81)$$
$$C_2 = 1.8019377358 \qquad L_2 = 2.0$$
$$C_3 = 1.8019377358 \qquad L_3 = 1.2469796037$$
$$C_4 = 0.4450418679$$

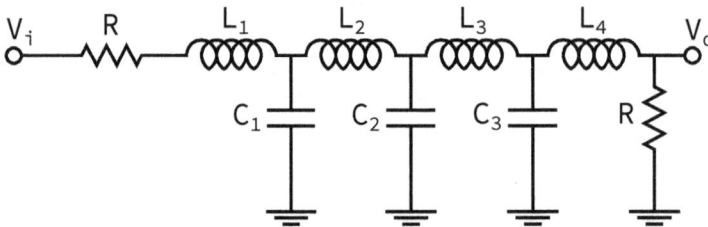

Figure 48: Seventh order low pass Butterworth with source and load impedance, variation 2.

$$L_1 = 0.4450418679 \qquad C_1 = 1.2469796037 \qquad (82)$$
$$L_2 = 1.8019377358 \qquad C_2 = 2.0$$
$$L_3 = 1.8019377358 \qquad C_3 = 1.2469796037$$
$$L_4 = 0.4450418679$$

Eighth Order Low Pass Butterworth

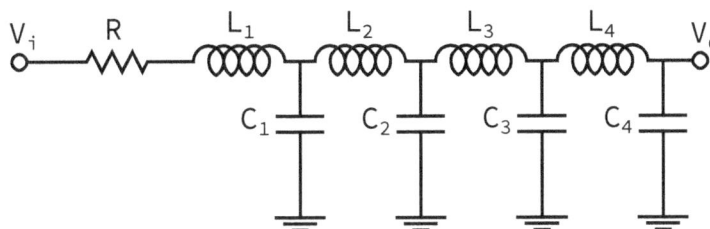

Figure 49: Eighth order low pass Butterworth with source impedance.

$$L_1 = 0.1950903220 \qquad C_1 = 0.5775519970 \qquad (83)$$
$$L_2 = 0.9370517338 \qquad C_2 = 1.2588210284$$
$$L_3 = 1.5283185530 \qquad C_3 = 1.7287352940$$
$$L_4 = 1.8246414248 \qquad C_4 = 1.5607225761$$

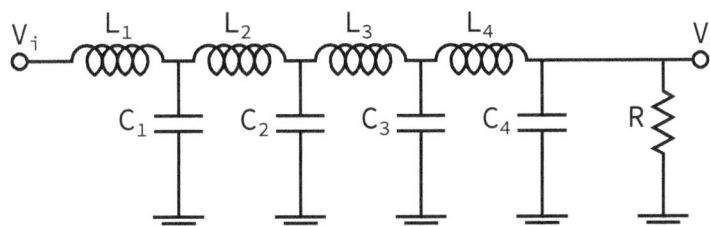

Figure 50: Eighth order low pass Butterworth with load impedance.

$$L_1 = 1.5607225761 \qquad C_1 = 1.8246414248 \qquad (84)$$
$$L_2 = 1.7287352940 \qquad C_2 = 1.5283185530$$
$$L_3 = 1.2588210284 \qquad C_3 = 0.9370517338$$
$$L_4 = 0.5775519970 \qquad C_4 = 0.1950903220$$

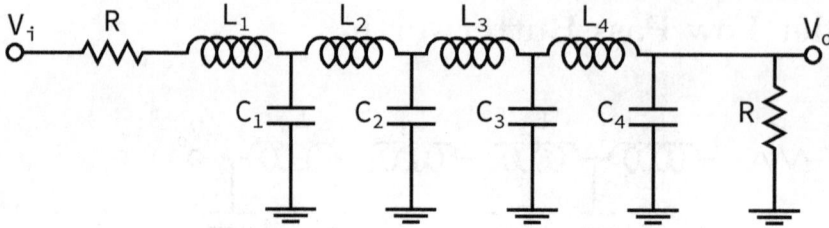

Figure 51: Eighth order low pass Butterworth with source and load impedance, variation 1.

$$L_1 = 0.3901806440 \qquad\qquad C_1 = 1.1111404660 \qquad\qquad (85)$$
$$L_2 = 1.6629392246 \qquad\qquad C_2 = 1.9615705608$$
$$L_3 = 1.9615705608 \qquad\qquad C_3 = 1.6629392246$$
$$L_4 = 1.1111404660 \qquad\qquad C_4 = 0.3901806440$$

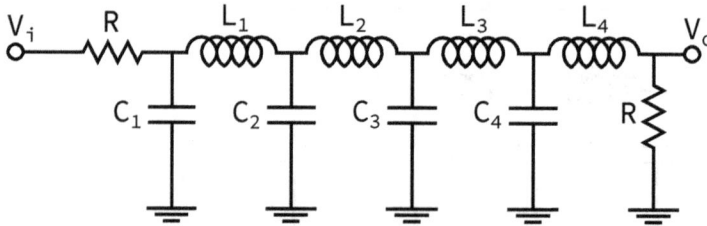

Figure 52: Eighth order low pass Butterworth with source and load impedance, variation 2.

$$C_1 = 0.3901806440 \qquad\qquad L_1 = 1.1111404660 \qquad\qquad (86)$$
$$C_2 = 1.6629392246 \qquad\qquad L_2 = 1.9615705608$$
$$C_3 = 1.9615705608 \qquad\qquad L_3 = 1.6629392246$$
$$C_4 = 1.1111404660 \qquad\qquad L_4 = 0.3901806440$$

Ninth Order Low Pass Butterworth

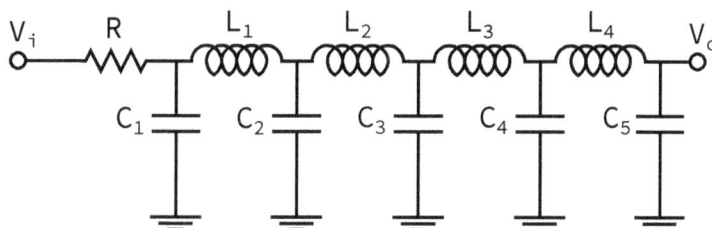

Figure 53: Ninth order low pass Butterworth with source impedance.

$$
\begin{array}{lll}
C_1 = 0.1736481777 & L_1 = 0.5155456021 & (87) \\
C_2 = 0.8413665688 & L_2 = 1.1407573299 & \\
C_3 = 1.4037333413 & L_3 = 1.6201909710 & \\
C_4 = 1.7771887964 & L_4 = 1.8424131932 & \\
C_5 = 1.5628335990 & &
\end{array}
$$

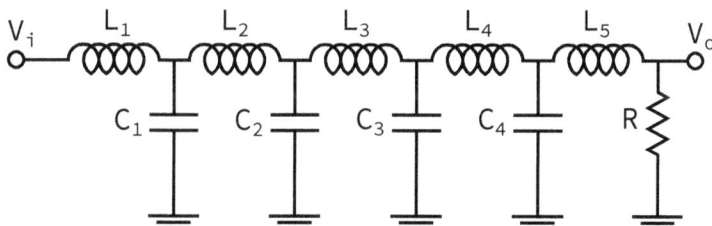

Figure 54: Ninth order low pass Butterworth with load impedance.

$$
\begin{array}{lll}
L_1 = 1.5628335990 & C_1 = 1.8424131932 & (88) \\
L_2 = 1.7771887964 & C_2 = 1.6201909710 & \\
L_3 = 1.4037333413 & C_3 = 1.1407573299 & \\
L_4 = 0.8413665688 & C_4 = 0.5155456021 & \\
L_5 = 0.1736481777 & &
\end{array}
$$

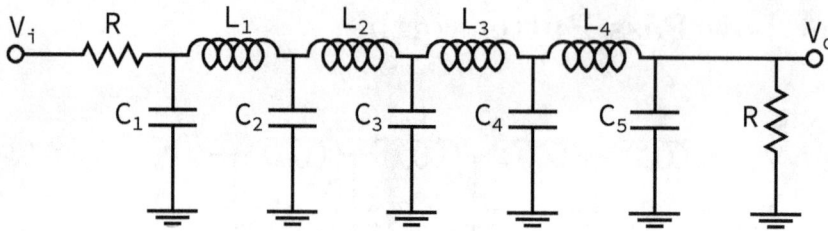

Figure 55: Ninth order low pass Butterworth with source and load impedance, variation 1.

$$C_1 = 0.3472963553 \qquad\qquad L_1 = 1.0 \tag{89}$$

$$C_2 = 1.5320888862 \qquad\qquad L_2 = 1.8793852416$$

$$C_3 = 2.0 \qquad\qquad L_3 = 1.8793852416$$

$$C_4 = 1.5320888862 \qquad\qquad L_4 = 1.0$$

$$C_5 = 0.3472963553$$

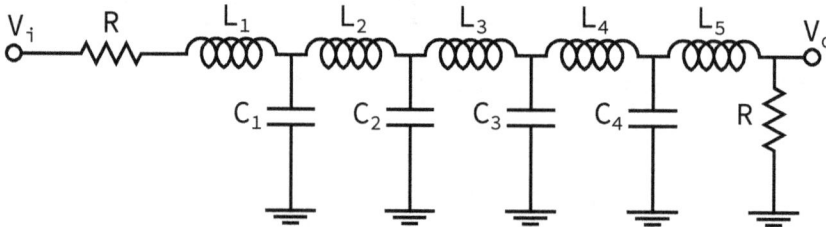

Figure 56: Ninth order low pass Butterworth with source and load impedance, variation 2.

$$L_1 = 0.3472963553 \qquad\qquad C_1 = 1.0 \tag{90}$$

$$L_2 = 1.5320888862 \qquad\qquad C_2 = 1.8793852416$$

$$L_3 = 2.0 \qquad\qquad C_3 = 1.8793852416$$

$$L_4 = 1.5320888862 \qquad\qquad C_4 = 1.0$$

$$L_5 = 0.3472963553$$

Tenth Order Low Pass Butterworth

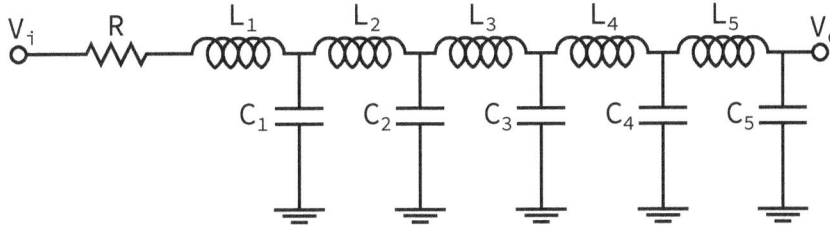

Figure 57: Tenth order low pass Butterworth with source impedance.

$$
\begin{aligned}
L_1 &= 0.1564344650 & C_1 &= 0.4653791379 & (91) \\
L_2 &= 0.7626270496 & C_2 &= 1.0406193876 \\
L_3 &= 1.2920925103 & C_3 &= 1.5099975433 \\
L_4 &= 1.6868916940 & C_4 &= 1.8121125023 \\
L_5 &= 1.8551621804 & C_5 &= 1.5643446504
\end{aligned}
$$

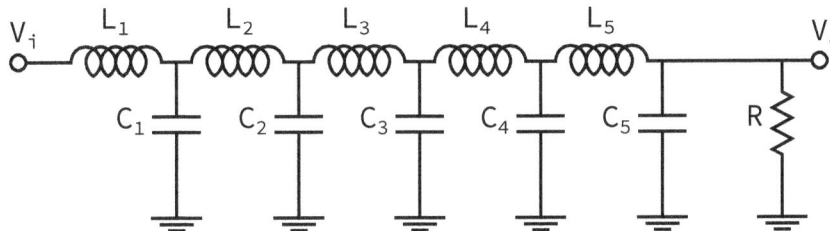

Figure 58: Tenth order low pass Butterworth with load impedance.

$$
\begin{aligned}
L_1 &= 1.5643446504 & C_1 &= 1.8551621804 & (92) \\
L_2 &= 1.8121125023 & C_2 &= 1.6868916940 \\
L_3 &= 1.5099975433 & C_3 &= 1.2920925103 \\
L_4 &= 1.0406193876 & C_4 &= 0.7626270496 \\
L_5 &= 0.4653791379 & C_5 &= 0.1564344650
\end{aligned}
$$

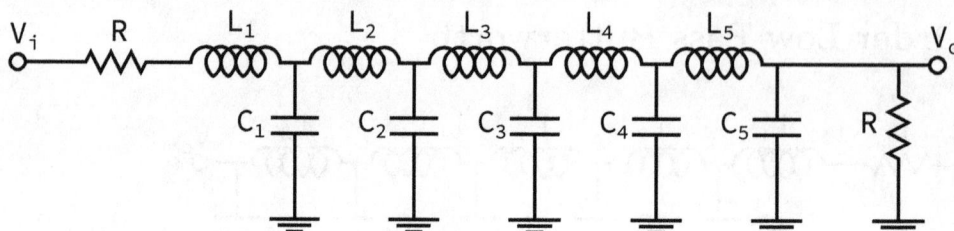

Figure 59: Tenth order low pass Butterworth with source and load impedance, variation 1.

$$L_1 = 0.3128689301 \qquad C_1 = 0.9079809995 \qquad (93)$$
$$L_2 = 1.4142135624 \qquad C_2 = 1.7820130484$$
$$L_3 = 1.9753766812 \qquad C_3 = 1.9753766812$$
$$L_4 = 1.7820130484 \qquad C_4 = 1.4142135624$$
$$L_5 = 0.9079809995 \qquad C_5 = 0.3128689301$$

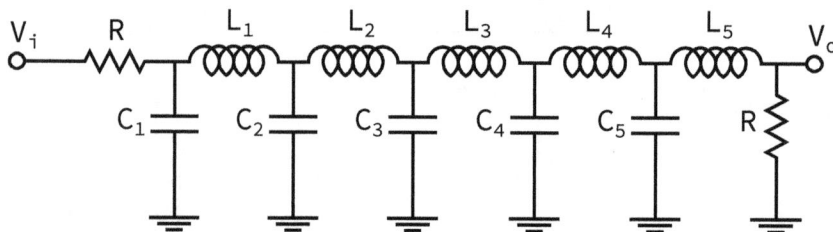

Figure 60: Tenth order low pass Butterworth with source and load impedance, variation 2.

$$C_1 = 0.3128689301 \qquad L_1 = 0.9079809995 \qquad (94)$$
$$C_2 = 1.4142135624 \qquad L_2 = 1.7820130484$$
$$C_3 = 1.9753766812 \qquad L_3 = 1.9753766812$$
$$C_4 = 1.7820130484 \qquad L_4 = 1.4142135624$$
$$C_5 = 0.9079809995 \qquad L_5 = 0.3128689301$$

High Pass Butterworth

A low pass Butterworth filter is transformed into a high pass filter with the following change in variables

$$s \to \frac{1}{s} \tag{95}$$

The change in variables corresponds to turning the series inductors in the low pass filter into capacitors and turning the shunt capacitors into inductors. If l is the inductance value in the low pass circuit and c is the capacitance value in the low pass circuit then the corresponding inductance and capacitance values in the high pass circuit are calculated as follows

$$C = \frac{1}{lR\omega_0} \qquad\qquad L = \frac{R}{\omega_0 c} \tag{96}$$

where R is the source or load resistance and ω_0 is the high pass cutoff frequency.

Circuit diagrams for the second through tenth order high pass filters are shown on the following pages. The given component values will produce a high pass filter with $R = 1$ and $\omega_0 = 1$. The frequency response for these filters is shown in figure 61.

The component values in these circuits can easily be converted to other values of R and ω_0 by scaling them as follows

$$L \to \frac{LR}{\omega_0} \qquad\qquad C \to \frac{C}{R\omega_0} \tag{97}$$

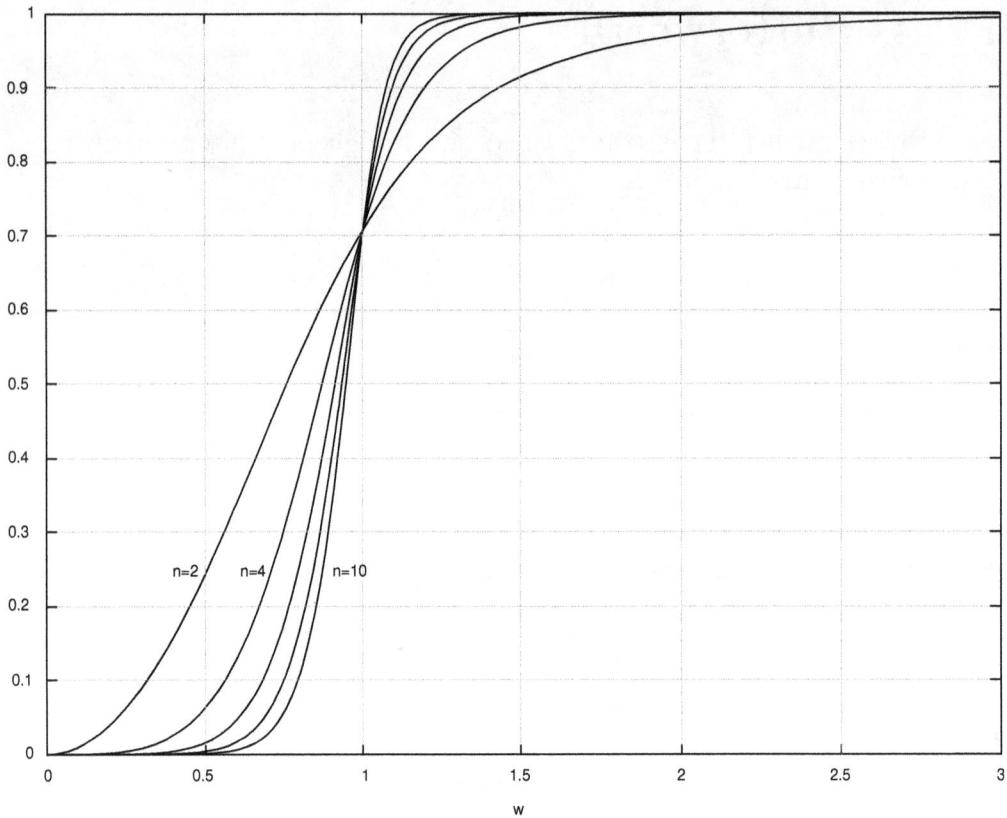

Figure 61: Frequency response for second, fourth, sixth and eighth order high pass filters.

For example if we want a fourth order high pass filter with cutoff at 100 kHz and a source resistance of 50 Ω then using the component values in equation 107 we get, for $R = 50\Omega$, and $\omega_0 = 2\pi \cdot 100000$,

$$
\begin{aligned}
L_1 &\to L_1 R/\omega_0 & &\approx 73.5\mu\text{H} \\
C_1 &\to C_1/R\omega_0 & &\approx 83.2\text{nF} \\
C_2 &\to C_2/R\omega_0 & &\approx 20.2\text{nF} \\
L_2 &\to L_2 R/\omega_0 & &\approx 52.0\mu\text{H}
\end{aligned}
\tag{98}
$$

The spice code and the frequency response simulation for the filter are shown below.

```
.title high pass 4th order with source resistance
V1 in 0 dc 0 ac 1
R1 in 1 50
C1 1 2 83.2n
L1 2 0 73.5u
C2 2 out 20.2n
L2 out 0 52.0u
.control
ac dec 1000 40000 200000
gnuplot newplot vm(out)
.endc
.end
```

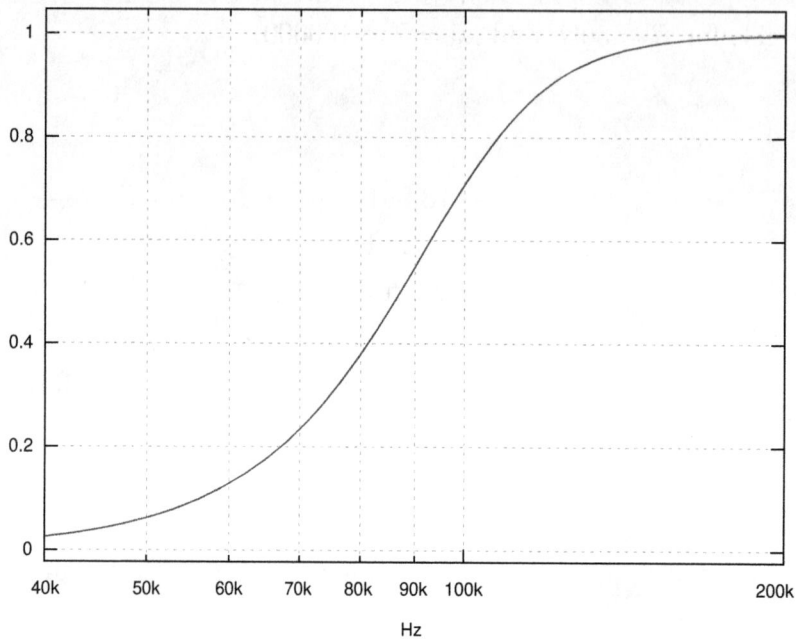

Figure 62: Spice generated frequency response for fourth order high pass filter with source resistance $R = 50\Omega$ and $L_1 = 73.5\mu H$, $C_1 = 83.2nF$, $C_2 = 20.2nF$ and $L_2 = 52.0\mu H$.

Second Order High Pass Butterworth

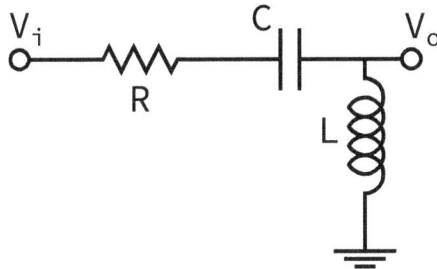

Figure 63: Second order high pass Butterworth with source impedance.

$$C = \sqrt{2} \qquad\qquad L = 1/\sqrt{2} \qquad\qquad (99)$$

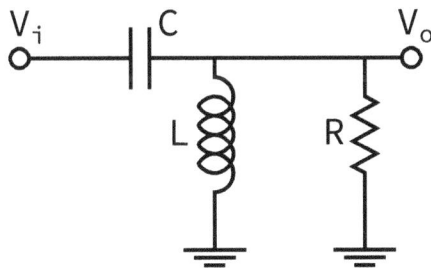

Figure 64: Second order high pass Butterworth with load impedance.

$$C = 1/\sqrt{2} \qquad\qquad L = \sqrt{2} \qquad\qquad (100)$$

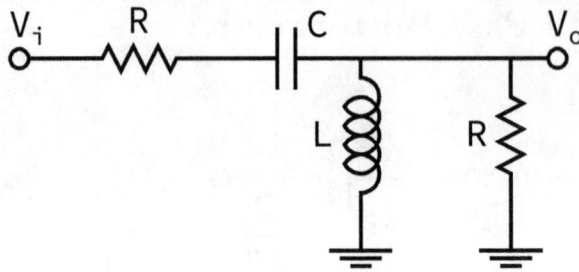

Figure 65: Second order high pass Butterworth with source and load impedance, variation 1.

$$C = 1/\sqrt{2} \qquad\qquad L = 1/\sqrt{2} \qquad\qquad (101)$$

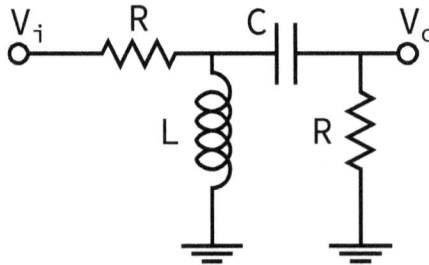

Figure 66: Second order high pass Butterworth with source and load impedance, variation 2.

$$L = 1/\sqrt{2} \qquad\qquad C = 1/\sqrt{2} \qquad\qquad (102)$$

Third Order High Pass Butterworth

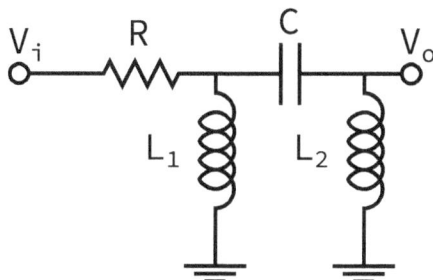

Figure 67: Third order high pass Butterworth with source impedance.

$$L_1 = 2 \qquad C = 3/4 \qquad L_2 = 2/3 \qquad (103)$$

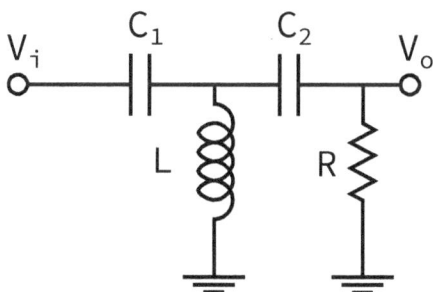

Figure 68: Third order high pass Butterworth with load impedance.

$$C_1 = 2/3 \qquad L = 3/4 \qquad C_2 = 2 \qquad (104)$$

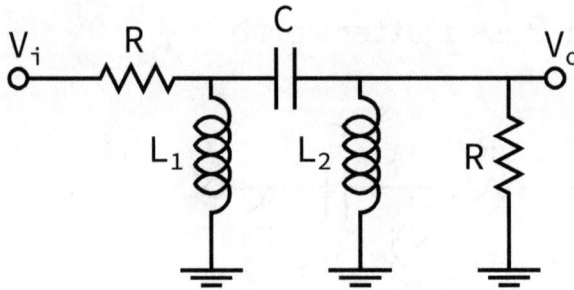

Figure 69: Third order high pass Butterworth with source and load impedance, variation 1.

$$L_1 = 1 \qquad C = 1/2 \qquad L_2 = 1 \qquad (105)$$

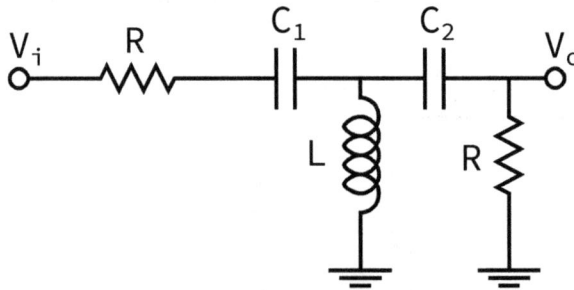

Figure 70: Third order high pass Butterworth with source and load impedance, variation 2.

$$C_1 = 1 \qquad L = 1/2 \qquad C_2 = 1 \qquad (106)$$

Fourth Order High Pass Butterworth

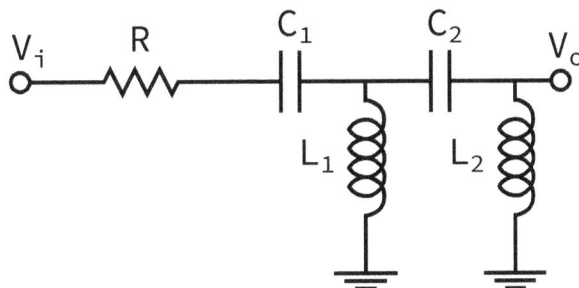

Figure 71: Fourth order high pass Butterworth with source impedance.

$$C_1 = 2.6131259298 \qquad L_1 = 0.9238795325 \qquad (107)$$
$$C_2 = 0.6340506711 \qquad L_2 = 0.6532814824$$

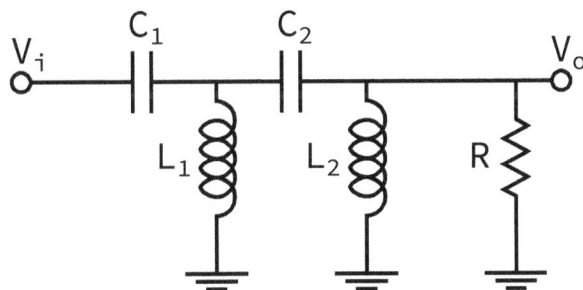

Figure 72: Fourth order high pass Butterworth with load impedance.

$$C_1 = 0.6532814824 \qquad L_1 = 0.6340506711 \qquad (108)$$
$$C_2 = 0.9238795325 \qquad L_2 = 2.6131259298$$

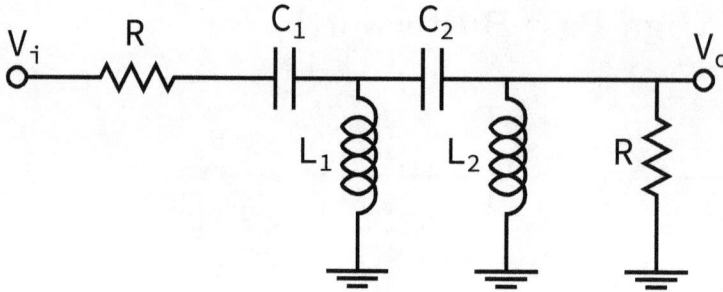

Figure 73: Fourth order high pass Butterworth with source and load impedance, variation 1.

$$C_1 = 1.3065629649 \qquad\qquad L_1 = 0.5411961001 \qquad\qquad (109)$$
$$C_2 = 0.5411961001 \qquad\qquad L_2 = 1.3065629649$$

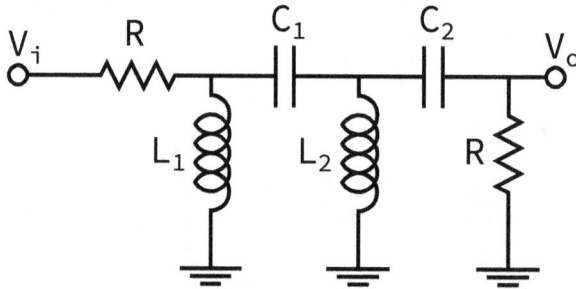

Figure 74: Fourth order high pass Butterworth with source and load impedance, variation 2.

$$L_1 = 1.3065629649 \qquad\qquad C_1 = 0.5411961001 \qquad\qquad (110)$$
$$L_2 = 0.5411961001 \qquad\qquad C_2 = 1.3065629649$$

Fifth Order High Pass Butterworth

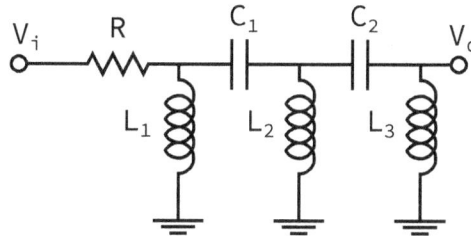

Figure 75: Fifth order high pass Butterworth with source impedance.

$$L_1 = 3.2360679775 \qquad C_1 = 1.1180339887 \qquad (111)$$
$$L_2 = 0.7236067977 \qquad C_2 = 0.5901699437$$
$$L_3 = 0.6472135955$$

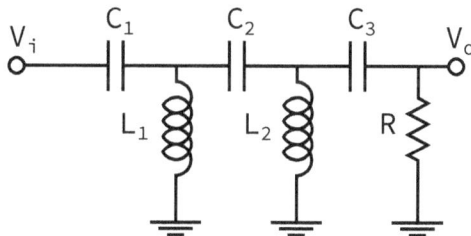

Figure 76: Fifth order high pass Butterworth with load impedance.

$$C_1 = 0.6472135955 \qquad L_1 = 0.5901699437 \qquad (112)$$
$$C_2 = 0.7236067977 \qquad L_2 = 1.1180339887$$
$$C_3 = 3.2360679775$$

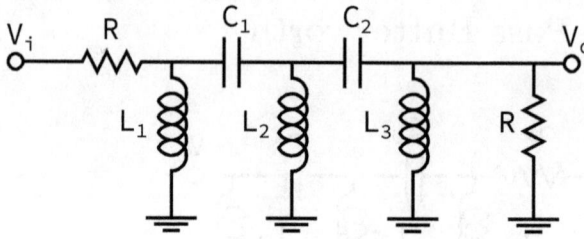

Figure 77: Fifth order high pass Butterworth with source and load impedance, variation 1.

$$L_1 = 1.6180339887 \qquad\qquad C_1 = 0.6180339887 \qquad\qquad (113)$$
$$L_2 = 0.5 \qquad\qquad\qquad\qquad C_2 = 0.6180339887$$
$$L_3 = 1.6180339887$$

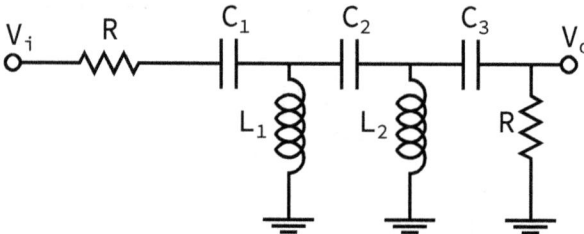

Figure 78: Fifth order high pass Butterworth with source and load impedance, variation 2.

$$C_1 = 1.6180339887 \qquad\qquad L_1 = 0.6180339887 \qquad\qquad (114)$$
$$C_2 = 0.5 \qquad\qquad\qquad\qquad L_2 = 0.6180339887$$
$$C_3 = 1.6180339887$$

Sixth Order High Pass Butterworth

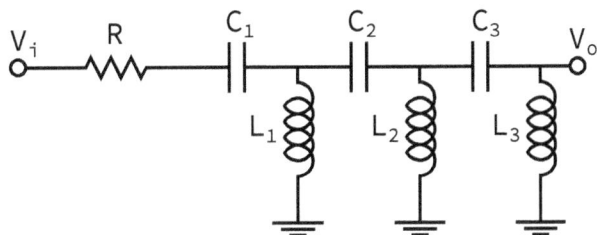

Figure 79: Sixth order high pass Butterworth with source impedance.

$$C_1 = 3.8637033052 \qquad\qquad L_1 = 1.3194792169 \qquad\qquad (115)$$
$$C_2 = 0.8322042494 \qquad\qquad L_2 = 0.6439505509$$
$$C_3 = 0.5684060729 \qquad\qquad L_3 = 0.6439505509$$

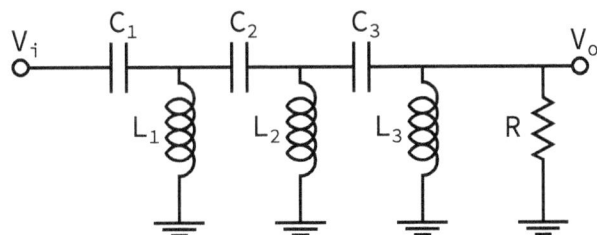

Figure 80: Sixth order high pass Butterworth with load impedance.

$$C_1 = 0.6439505509 \qquad\qquad L_1 = 0.5684060729 \qquad\qquad (116)$$
$$C_2 = 0.6439505509 \qquad\qquad L_2 = 0.8322042494$$
$$C_3 = 1.3194792169 \qquad\qquad L_3 = 3.8637033052$$

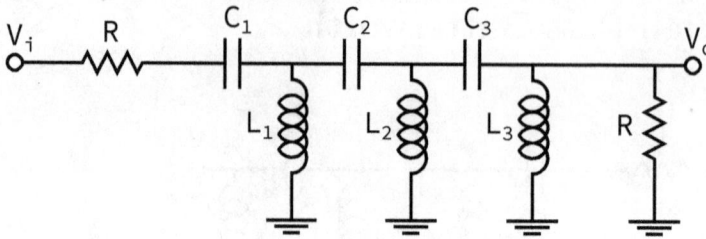

Figure 81: Sixth order high pass Butterworth with source and load impedance, variation 1.

$$C_1 = 1.9318516526 \qquad\qquad L_1 = 0.7071067812 \qquad\qquad (117)$$
$$C_2 = 0.5176380902 \qquad\qquad L_2 = 0.5176380902$$
$$C_3 = 0.7071067812 \qquad\qquad L_3 = 1.9318516526$$

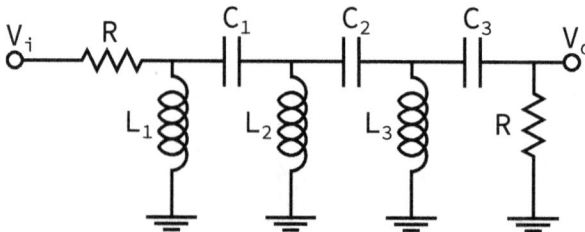

Figure 82: Sixth order high pass Butterworth with source and load impedance, variation 2.

$$L_1 = 1.9318516526 \qquad\qquad C_1 = 0.7071067812 \qquad\qquad (118)$$
$$L_2 = 0.5176380902 \qquad\qquad C_2 = 0.5176380902$$
$$L_3 = 0.7071067812 \qquad\qquad C_3 = 1.9318516526$$

Seventh Order High Pass Butterworth

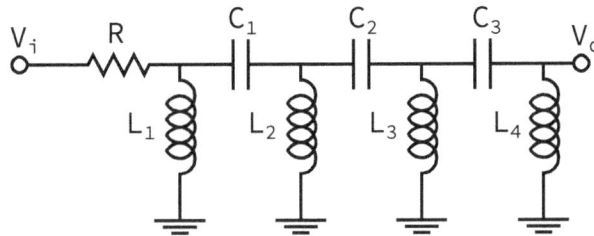

Figure 83: Seventh order high pass Butterworth with source impedance.

$$L_1 = 4.4939592074 \qquad C_1 = 1.5244586698 \qquad (119)$$
$$L_2 = 0.9479049164 \qquad C_2 = 0.7157341651$$
$$L_3 = 0.6028332183 \qquad C_3 = 0.5559176129$$
$$L_4 = 0.6419941725$$

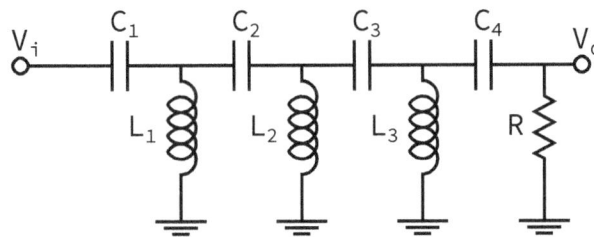

Figure 84: Seventh order high pass Butterworth with load impedance.

$$C_1 = 0.6419941725 \qquad L_1 = 0.5559176129 \qquad (120)$$
$$C_2 = 0.6028332183 \qquad L_2 = 0.7157341651$$
$$C_3 = 0.9479049164 \qquad L_3 = 1.5244586698$$
$$C_4 = 4.4939592074$$

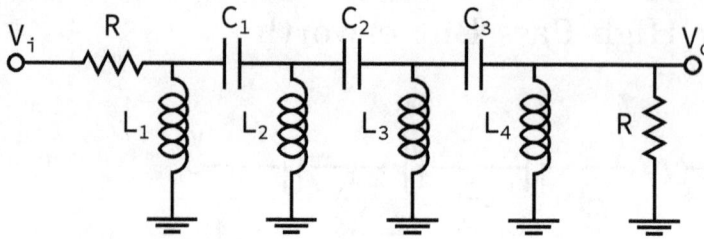

Figure 85: Seventh order high pass Butterworth with source and load impedance, variation 1.

$$L_1 = 2.2469796037 \qquad\qquad C_1 = 0.8019377358 \qquad\qquad (121)$$
$$L_2 = 0.5549581321 \qquad\qquad C_2 = 0.5$$
$$L_3 = 0.5549581321 \qquad\qquad C_3 = 0.8019377358$$
$$L_4 = 2.2469796037$$

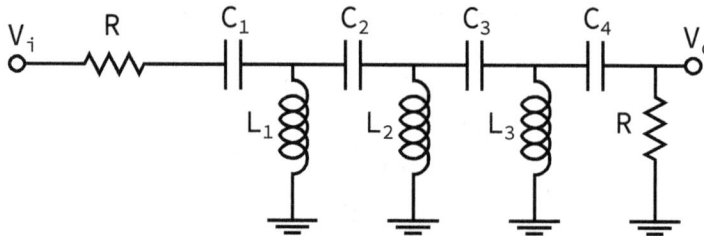

Figure 86: Seventh order high pass Butterworth with source and load impedance, variation 2.

$$C_1 = 2.2469796037 \qquad\qquad L_1 = 0.8019377358 \qquad\qquad (122)$$
$$C_2 = 0.5549581321 \qquad\qquad L_2 = 0.5$$
$$C_3 = 0.5549581321 \qquad\qquad L_3 = 0.8019377358$$
$$C_4 = 2.2469796037$$

Eighth Order High Pass Butterworth

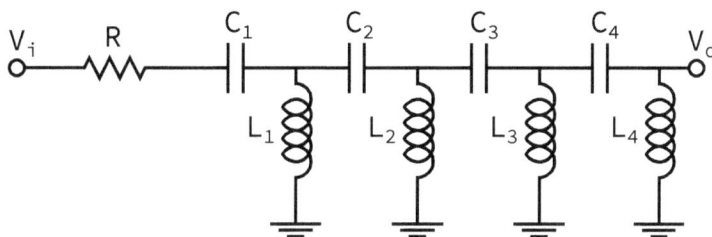

Figure 87: Eighth order high pass Butterworth with source impedance.

$$C_1 = 5.1258308955 \qquad\qquad L_1 = 1.7314458354 \qquad\qquad (123)$$
$$C_2 = 1.0671769380 \qquad\qquad L_2 = 0.7943941017$$
$$C_3 = 0.6543138523 \qquad\qquad L_3 = 0.5784575600$$
$$C_4 = 0.5480528867 \qquad\qquad L_4 = 0.6407288619$$

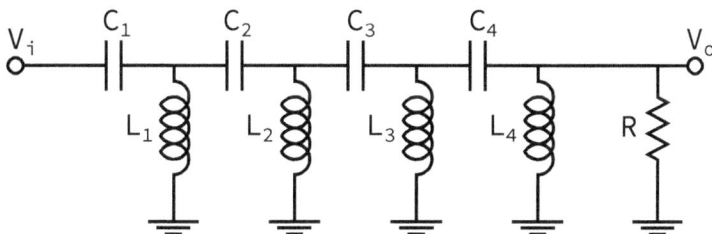

Figure 88: Eighth order high pass Butterworth with load impedance.

$$C_1 = 0.6407288619 \qquad\qquad L_1 = 0.5480528867 \qquad\qquad (124)$$
$$C_2 = 0.5784575600 \qquad\qquad L_2 = 0.6543138523$$
$$C_3 = 0.7943941017 \qquad\qquad L_3 = 1.0671769380$$
$$C_4 = 1.7314458354 \qquad\qquad L_4 = 5.1258308955$$

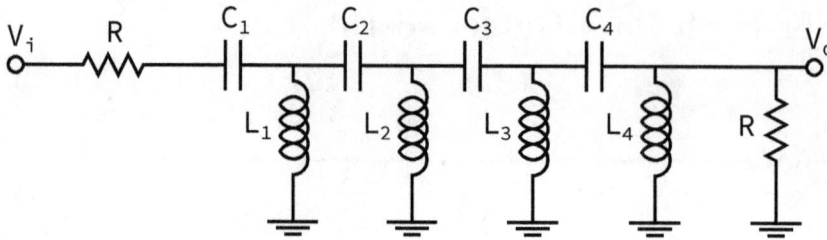

Figure 89: Eighth order high pass Butterworth with source and load impedance, variation 1.

$$C_1 = 2.5629154477 \qquad L_1 = 0.8999762231 \qquad (125)$$
$$C_2 = 0.6013448869 \qquad L_2 = 0.5097955791$$
$$C_3 = 0.5097955791 \qquad L_3 = 0.6013448869$$
$$C_4 = 0.8999762231 \qquad L_4 = 2.5629154477$$

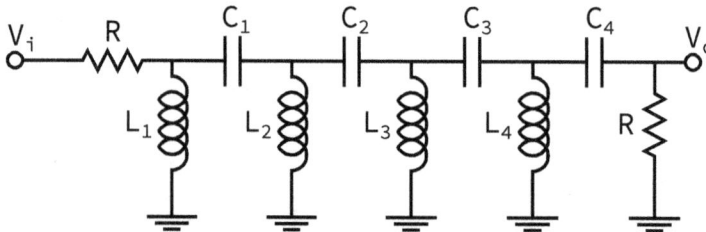

Figure 90: Eighth order high pass Butterworth with source and load impedance, variation 2.

$$L_1 = 2.5629154477 \qquad C_1 = 0.8999762231 \qquad (126)$$
$$L_2 = 0.6013448869 \qquad C_2 = 0.5097955791$$
$$L_3 = 0.5097955791 \qquad C_3 = 0.6013448869$$
$$L_4 = 0.8999762231 \qquad C_4 = 2.5629154477$$

Ninth Order High Pass Butterworth

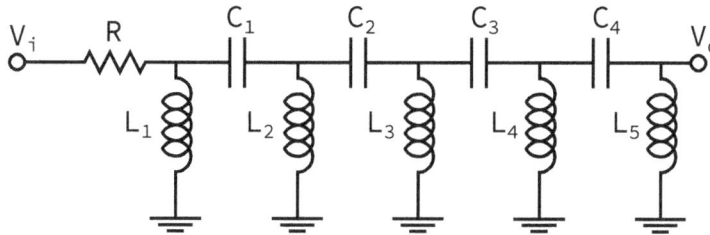

Figure 91: Ninth order high pass Butterworth with source impedance.

$L_1 = 5.7587704831$ $C_1 = 1.9396926208$ (127)

$L_2 = 1.1885425890$ $C_2 = 0.8766106286$

$L_3 = 0.7123860142$ $C_3 = 0.6172111917$

$L_4 = 0.5626864192$ $C_4 = 0.5427664129$

$L_5 = 1.5628335990$

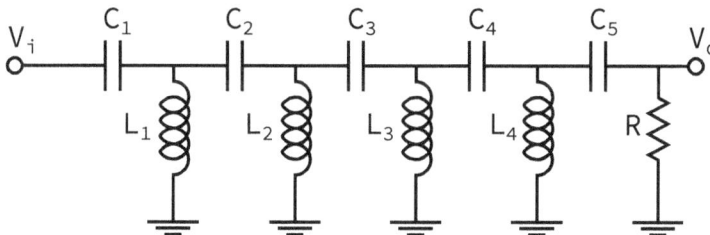

Figure 92: Ninth order high pass Butterworth with load impedance.

$C_1 = 0.6398633870$ $L_1 = 0.5427664129$ (128)

$C_2 = 0.5626864192$ $L_2 = 0.6172111917$

$C_3 = 0.7123860142$ $L_3 = 0.8766106286$

$C_4 = 1.1885425890$ $L_4 = 1.9396926208$

$C_5 = 5.7587704831$

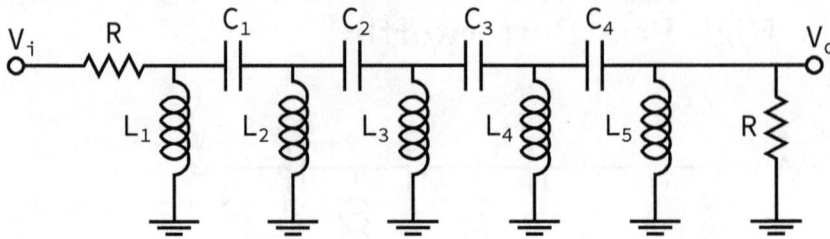

Figure 93: Ninth order high pass Butterworth with source and load impedance, variation 1.

$$
\begin{aligned}
&L_1 = 2.8793852416 && C_1 = 1.0 && (129) \\
&L_2 = 0.6527036447 && C_2 = 0.5320888862 \\
&L_3 = 0.5 && C_3 = 0.5320888862 \\
&L_4 = 0.6527036447 && C_4 = 1.0 \\
&L_5 = 0.3472963553
\end{aligned}
$$

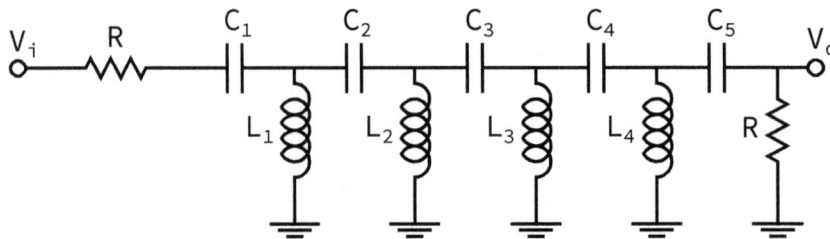

Figure 94: Ninth order high pass Butterworth with source and load impedance, variation 2.

$$
\begin{aligned}
&C_1 = 2.8793852416 && L_1 = 1.0 && (130) \\
&C_2 = 0.6527036447 && L_2 = 0.5320888862 \\
&C_3 = 0.5 && L_3 = 0.5320888862 \\
&C_4 = 0.6527036447 && L_4 = 1.0 \\
&C_5 = 2.8793852416
\end{aligned}
$$

Tenth Order High Pass Butterworth

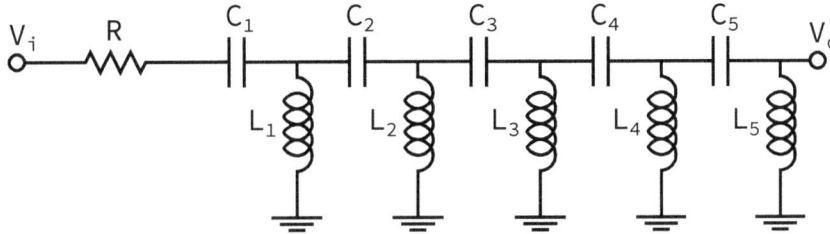

Figure 95: Tenth order high pass Butterworth with source impedance.

$$
\begin{aligned}
C_1 &= 6.3924532215 & L_1 &= 2.1487856215 & (131)\\
C_2 &= 1.3112569250 & L_2 &= 0.9609661438 \\
C_3 &= 0.7739383922 & L_3 &= 0.6622527331 \\
C_4 &= 0.5928062860 & L_4 &= 0.5518421173 \\
C_5 &= 0.5390364306 & L_5 &= 0.6392453221
\end{aligned}
$$

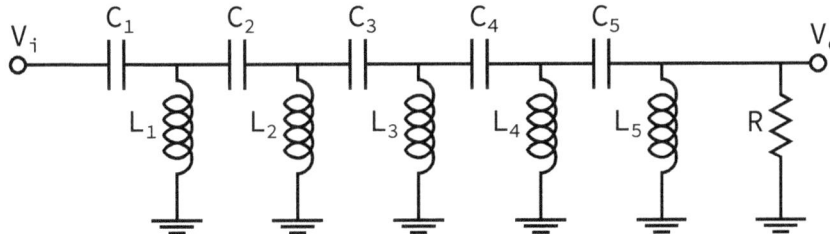

Figure 96: Tenth order high pass Butterworth with load impedance.

$$
\begin{aligned}
C_1 &= 0.6392453221 & L_1 &= 0.5390364306 & (132)\\
C_2 &= 0.5518421173 & L_2 &= 0.5928062860 \\
C_3 &= 0.6622527331 & L_3 &= 0.7739383922 \\
C_4 &= 0.9609661438 & L_4 &= 1.3112569250 \\
C_5 &= 2.1487856215 & L_5 &= 6.3924532215
\end{aligned}
$$

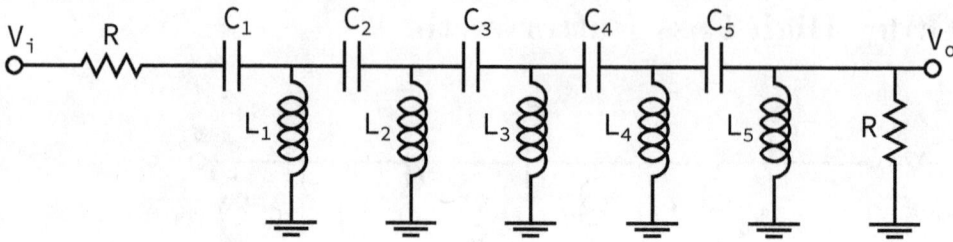

Figure 97: Tenth order high pass Butterworth with source and load impedance, variation 1.

$$C_1 = 3.1962266107 \qquad L_1 = 1.1013446323 \qquad (133)$$
$$C_2 = 0.7071067812 \qquad L_2 = 0.5611631188$$
$$C_3 = 0.5062325629 \qquad L_3 = 0.5062325629$$
$$C_4 = 0.5611631188 \qquad L_4 = 0.7071067812$$
$$C_5 = 1.1013446323 \qquad L_5 = 3.1962266107$$

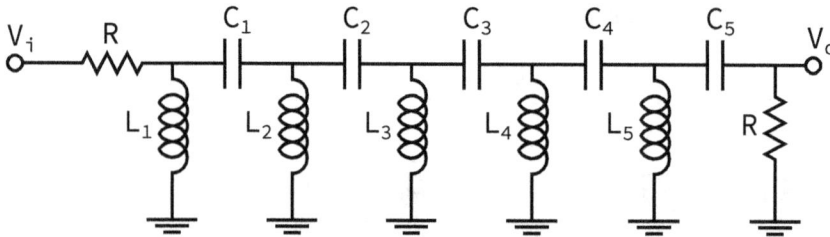

Figure 98: Tenth order high pass Butterworth with source and load impedance, variation 2.

$$L_1 = 3.1962266107 \qquad C_1 = 1.1013446323 \qquad (134)$$
$$L_2 = 0.7071067812 \qquad C_2 = 0.5611631188$$
$$L_3 = 0.5062325629 \qquad C_3 = 0.5062325629$$
$$L_4 = 0.5611631188 \qquad C_4 = 0.7071067812$$
$$L_5 = 1.1013446323 \qquad C_5 = 3.1962266107$$

Band Pass Butterworth

A low pass Butterworth filter is transformed into a band pass filter with the following change in variables

$$s \to \frac{\omega_0}{\Delta}\left(s + \frac{1}{s}\right) \tag{135}$$

where Δ is the bandwidth given by

$$\Delta = \omega_2 - \omega_1 \tag{136}$$

with $\omega_1 = 2\pi f_1$ and $\omega_2 = 2\pi f_2$ being the lower and upper cutoff frequencies respectively. The center frequency is ω_0 which is where the frequency response peaks. It is the geometric mean of the two cutoff frequencies

$$\omega_0 = \sqrt{\omega_1 \omega_2} \tag{137}$$

For small bandwidth this is approximately equal to the arithmetic mean of ω_1 and ω_2.

If Δ and ω_0 are specified then the lower and upper cutoff frequencies can be calculated using the following formulas

$$\omega_1 = \frac{\Delta}{2}\left(\sqrt{\left(\frac{2\omega_0}{\Delta}\right)^2 + 1} - 1\right)$$

$$\omega_2 = \frac{\Delta}{2}\left(\sqrt{\left(\frac{2\omega_0}{\Delta}\right)^2 + 1} + 1\right) \tag{138}$$

The change in variables corresponds to turning the series inductors in the low pass filter into a series resonant inductor and capacitor as shown in figure 99. If l is the inductance value in the low pass circuit then the inductance and capacitance values in the band pass circuit are related to it as follows

$$L = \frac{lR}{\Delta} \qquad\qquad C = \frac{\Delta}{R\omega_0^2 l} \qquad\qquad (139)$$

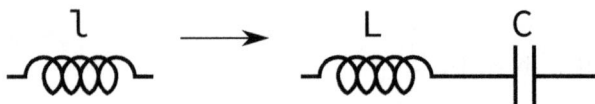

Figure 99: Series inductor transformation.

The shunt capacitors in the low pass filter are turned into a parallel resonant capacitor and inductor as shown in figure 100. If c is the capacitance value in the low pass circuit then the inductance and capacitance values in the band pass circuit are related to it as follows

$$C = \frac{c}{R\Delta} \qquad\qquad L = \frac{R\Delta}{\omega_0^2 c} \qquad\qquad (140)$$

R in these transform equations is the source or load resistance in the band pass circuit.

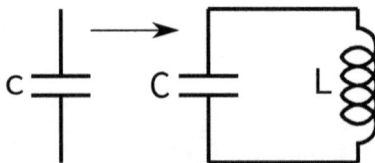

Figure 100: Shunt capacitor transformation.

Circuit diagrams for the second, fourth, sixth and eighth order band pass filters are shown on the following pages. The given component values will produce a band pass filter with $R = 1$, $\omega_0 = 1$, $\Delta = 1$, $\omega_1 = 1/\phi$ and $\omega_2 = \phi$ where $\phi = (\sqrt{5} + 1)/2$ is the golden ratio. The frequency response for these filters is shown in figure 101.

The component values in these circuits can easily be converted to other values of R, ω_0 and Δ by scaling the values in the series branches as follows

$$L \to \frac{LR}{\Delta} \qquad\qquad C \to \frac{C\Delta}{R\omega_0^2} \tag{141}$$

The values in the shunt branches are scaled as follows

$$C \to \frac{C}{R\Delta} \qquad\qquad L \to \frac{LR\Delta}{\omega_0^2} \tag{142}$$

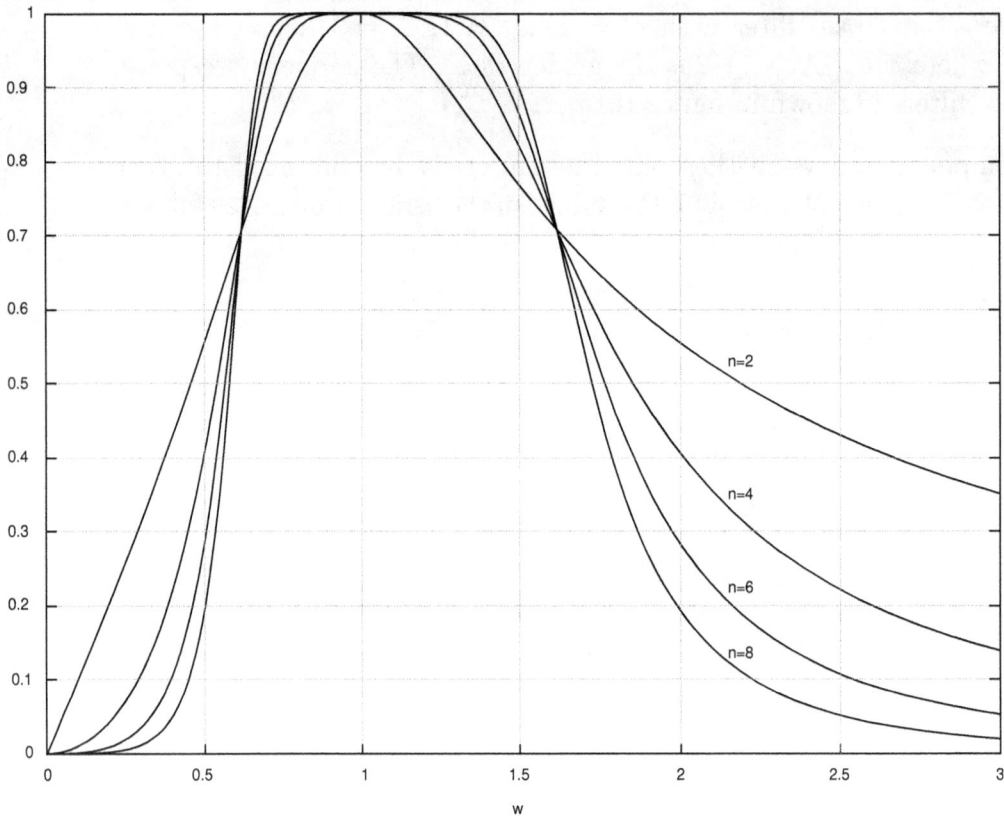

Figure 101: Frequency response for second, fourth, sixth and eighth order band pass filters.

For example if we want a fourth order band pass filter centered at 1 MHz with a bandwidth of 100 kHz and a source resistance of 50 Ω then using the component values in equation 148 we get, for $R = 50\Omega$, $\Delta = 2\pi \cdot 100000$

and $\omega_0 = 2\pi \cdot 1000000$,

$$L_1 \to L_1 R/\Delta \qquad\qquad \approx 56.3\mu H$$
$$C_1 \to C_1 \Delta/R\omega_0^2 \qquad\qquad \approx 0.45nF$$
$$C_2 \to C_2/R\Delta \qquad\qquad \approx 45nF$$
$$L_2 \to L_2 R\Delta/\omega_0^2 \qquad\qquad \approx 0.563\mu H$$

$$(143)$$

The upper and lower cutoff frequencies for the filter are calculated using eq. 138. They are $f_1 = 951249$ Hz and $f_2 = 1051249$ Hz.

The spice code and the frequency response simulation for the filter are shown below.

```
.title band pass 4th order with source resistance
V1 in 0 dc 0 ac 1
R1 in 1 50
L1 1 2 56.3u
C1 2 out 0.45n
L2 out 0 0.563u
C2 out 0 45n
.control
ac dec 1000 500000 2000000
gnuplot newplot vm(out)
.endc
.end
```

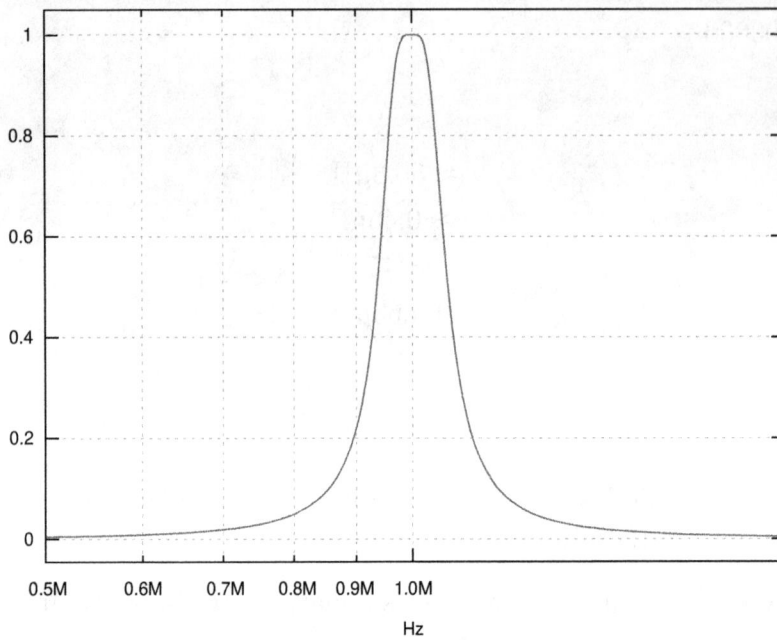

Figure 102: Spice generated frequency response for fourth order band pass filter with source resistance $R = 50\Omega$ and $L_1 = 56.3\mu$H, $C_1 = 0.45$nF, $C_2 = 45$nF and $L_2 = 0.563\mu$H.

Second Order Band Pass Butterworth

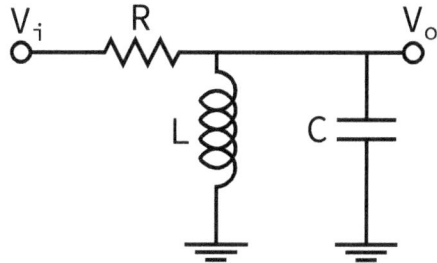

Figure 103: Second order band pass Butterworth with source impedance.

$$R = 1 \qquad C = 1 \qquad L = 1 \qquad (144)$$

These values produce a band pass filter centered at $\omega_0 = 1$ or $f_0 = 1/2\pi$ with a bandwidth of $\Delta = 1$. Lower and upper cutoff frequencies are at $\omega_1 = 1/\phi$, $\omega_2 = \phi$ where ϕ is the golden ratio, $\phi = (\sqrt{5} + 1)/2$.

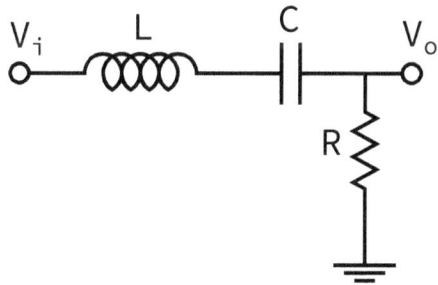

Figure 104: Second order band pass Butterworth with load impedance.

$$R = 1 \qquad C = 1 \qquad L = 1 \qquad (145)$$

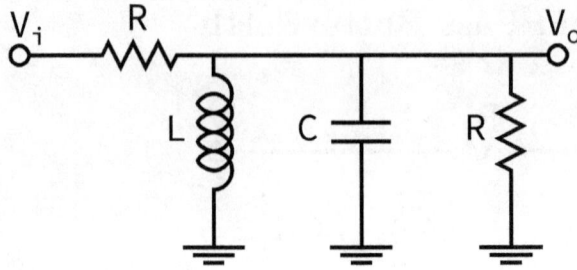

Figure 105: Second order band pass Butterworth with source and load impedance, variation 1.

$$R = 1 \qquad\qquad C = 2 \qquad\qquad L = 1/2 \qquad\qquad\qquad (146)$$

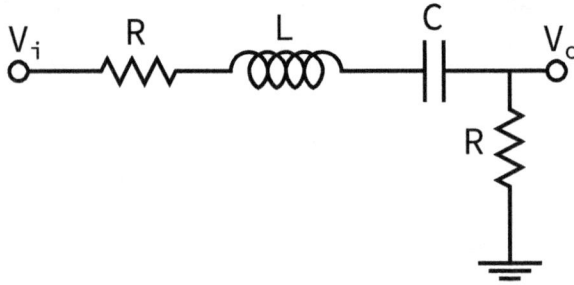

Figure 106: Second order band pass Butterworth with source and load impedance, variation 2.

$$R = 1 \qquad\qquad L = 2 \qquad\qquad C = 1/2 \qquad\qquad\qquad (147)$$

Fourth Order Band Pass Butterworth

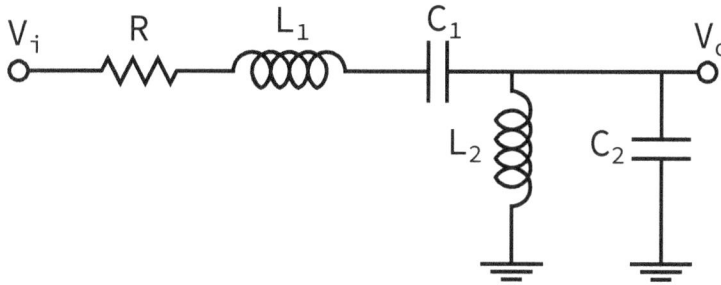

Figure 107: Fourth order band pass Butterworth with source impedance.

$$L_1 = 1/\sqrt{2} \qquad C_1 = \sqrt{2} \qquad L_2 = 1/\sqrt{2} \qquad C_2 = \sqrt{2} \tag{148}$$

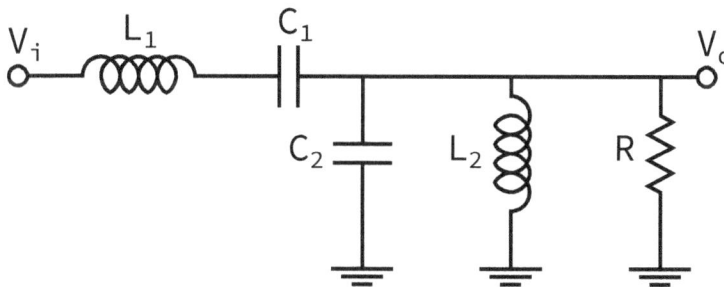

Figure 108: Fourth order band pass Butterworth with load impedance.

$$L_1 = \sqrt{2} \qquad C_1 = 1/\sqrt{2} \qquad L_2 = \sqrt{2} \qquad C_2 = 1/\sqrt{2} \tag{149}$$

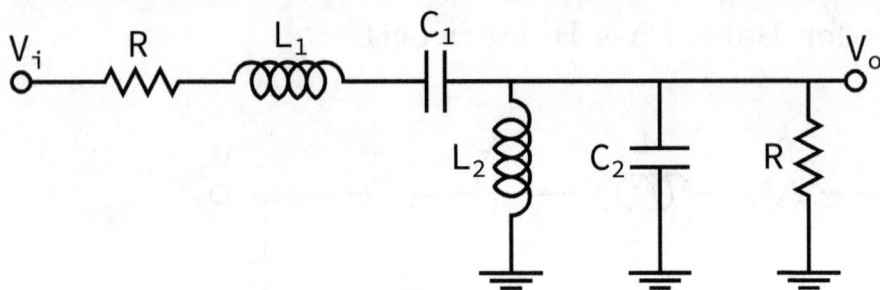

Figure 109: Fourth order band pass Butterworth with source and load impedance, variation 1.

$$L_1 = \sqrt{2} \qquad C_1 = 1/\sqrt{2} \qquad L_2 = 1/\sqrt{2} \qquad C_2 = \sqrt{2} \qquad (150)$$

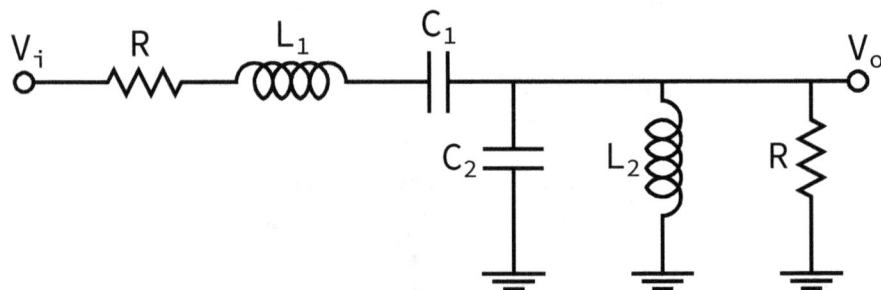

Figure 110: Fourth order band pass Butterworth with source and load impedance, variation 2.

$$L_1 = 1/\sqrt{2} \qquad C_1 = \sqrt{2} \qquad L_2 = \sqrt{2} \qquad C_2 = 1/\sqrt{2} \qquad (151)$$

Sixth Order Band Pass Butterworth

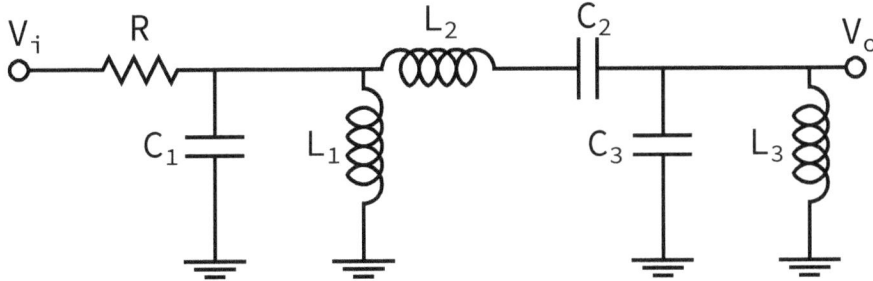

Figure 111: Sixth order band pass Butterworth with source impedance.

$$C_1 = 1/2 \quad L_1 = 2 \quad C_2 = 3/4 \quad L_2 = 4/3 \quad C_3 = 3/2 \quad L_3 = 2/3 \tag{152}$$

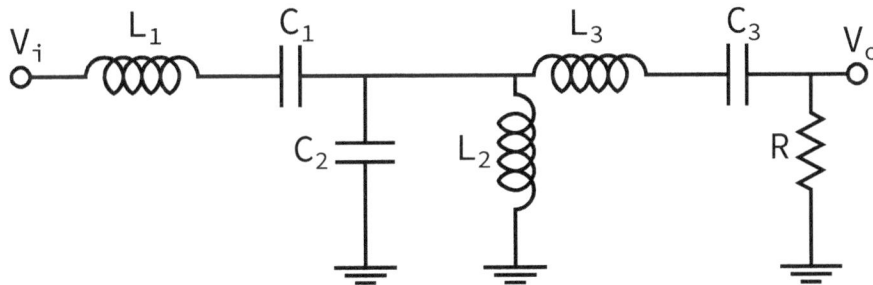

Figure 112: Sixth order band pass Butterworth with load impedance.

$$L_1 = 3/2 \quad C_1 = 2/3 \quad C_2 = 4/3 \quad L_2 = 3/4 \quad L_3 = 1/2 \quad C_3 = 2 \tag{153}$$

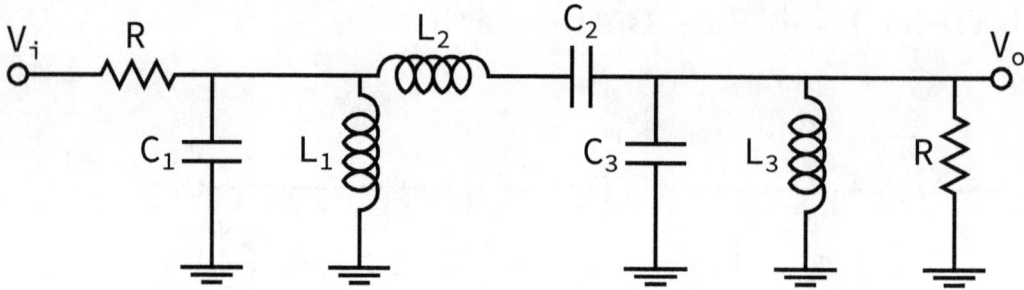

Figure 113: Sixth order band pass Butterworth with source and load impedance, variation 1.

$$C_1 = 1 \quad L_1 = 1 \quad C_2 = 1/2 \quad L_2 = 2 \quad C_3 = 1 \quad L_3 = 1 \qquad (154)$$

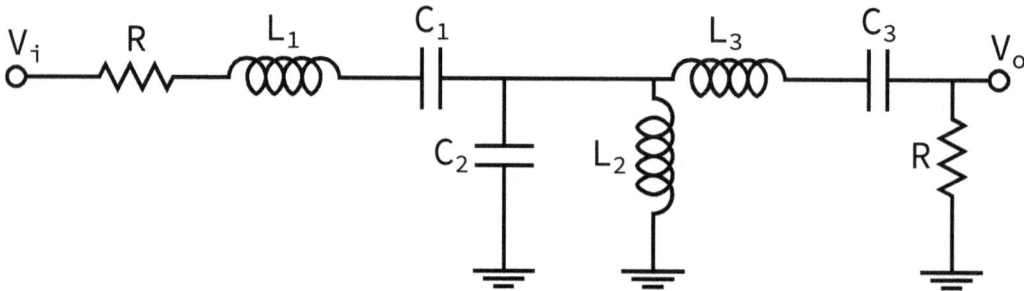

Figure 114: Sixth order band pass Butterworth with source and load impedance, variation 2.

$$L_1 = 1 \quad C_1 = 1 \quad C_2 = 2 \quad L_2 = 1/2 \quad L_3 = 1 \quad C_3 = 1 \qquad (155)$$

Eighth Order Band Pass Butterworth

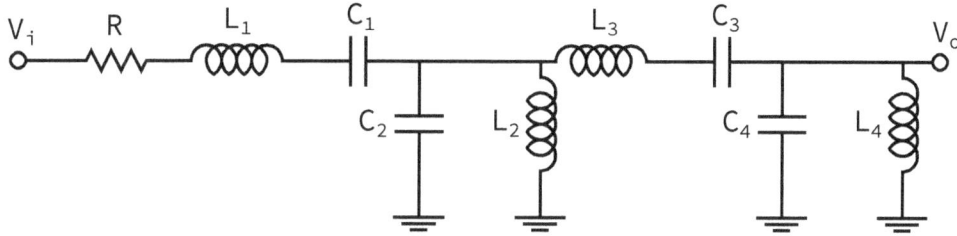

Figure 115: Eighth order band pass Butterworth with source impedance.

$L_1 = 0.3826834324$ $C_1 = 2.6131259298$

$C_2 = 1.0823922003$ $L_2 = 0.9238795325$

$L_3 = 1.5771610149$ $C_3 = 0.6340506711$

$C_4 = 1.5307337295$ $L_4 = 0.6532814824$ (156)

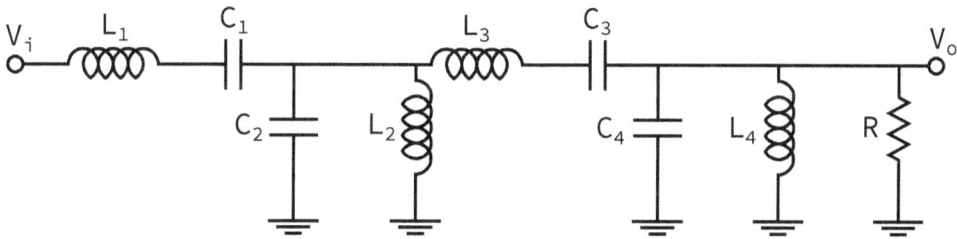

Figure 116: Eighth order band pass Butterworth with load impedance.

$L_1 = 1.5307337295$ $C_1 = 0.6532814824$

$C_2 = 1.5771610149$ $L_2 = 0.6340506711$

$L_3 = 1.0823922003$ $C_3 = 0.9238795325$

$C_4 = 0.3826834324$ $L_4 = 2.6131259298$ (157)

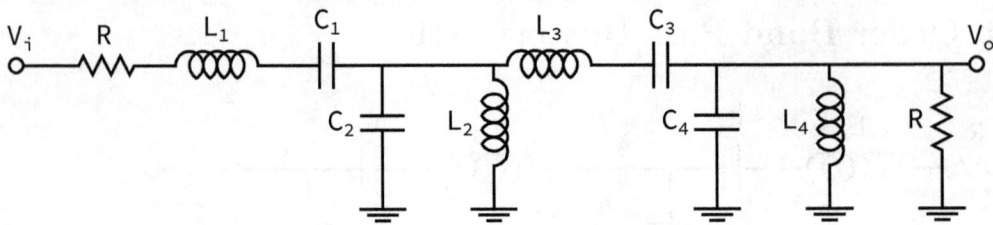

Figure 117: Eighth order band pass Butterworth with source and load impedance, variation 1.

$$L_1 = 0.7653668647 \qquad\qquad C_1 = 1.3065629649$$
$$C_2 = 1.8477590650 \qquad\qquad L_2 = 0.5411961001$$
$$L_3 = 1.8477590650 \qquad\qquad C_3 = 0.5411961001$$
$$C_4 = 0.7653668647 \qquad\qquad L_4 = 1.3065629649 \qquad (158)$$

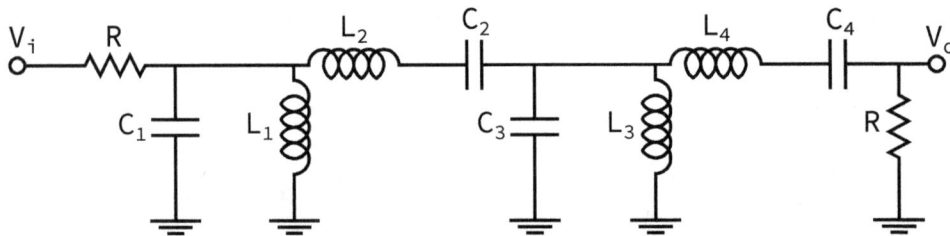

Figure 118: Eighth order band pass Butterworth with source and load impedance, variation 2.

$$C_1 = 0.7653668647 \qquad\qquad L_1 = 1.3065629649$$
$$L_2 = 1.8477590650 \qquad\qquad C_2 = 0.5411961001$$
$$C_3 = 1.8477590650 \qquad\qquad L_3 = 0.5411961001$$
$$L_4 = 0.7653668647 \qquad\qquad C_4 = 1.3065629649 \qquad (159)$$

Band Stop Butterworth

A low pass Butterworth filter is transformed into a band stop filter with the following change in variables

$$s \to \frac{\Delta}{\omega_0} \left(s + \frac{1}{s} \right)^{-1} \tag{160}$$

where Δ is the bandwidth given by

$$\Delta = \omega_2 - \omega_1 \tag{161}$$

with $\omega_1 = 2\pi f_1$ and $\omega_2 = 2\pi f_2$ being the lower and upper cutoff frequencies respectively. The center frequency is ω_0 which is where the frequency response has a minimum. It is the geometric mean of the two cutoff frequencies

$$\omega_0 = \sqrt{\omega_1 \omega_2} \tag{162}$$

For small bandwidth this is approximately equal to the arithmetic mean of ω_1 and ω_2.

If Δ and ω_0 are specified then the lower and upper cutoff frequencies can be calculated using the following formulas

$$\omega_1 = \frac{\Delta}{2} \left(\sqrt{\left(\frac{2\omega_0}{\Delta} \right)^2 + 1} - 1 \right)$$

$$\omega_2 = \frac{\Delta}{2} \left(\sqrt{\left(\frac{2\omega_0}{\Delta} \right)^2 + 1} + 1 \right) \tag{163}$$

The change in variables corresponds to turning the series inductors in the low pass filter into a parallel resonant inductor and capacitor as shown in figure 119. If l is the inductance value in the low pass circuit then the inductance and capacitance values in the band stop circuit are related to it as follows

$$L = \frac{lR\Delta}{\omega_0^2} \qquad\qquad C = \frac{1}{lR\Delta} \qquad\qquad (164)$$

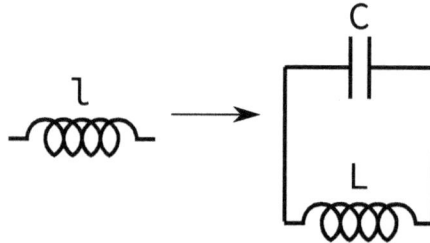

Figure 119: Series inductor transformation.

The shunt capacitors in the low pass filter are turned into a series resonant capacitor and inductor as shown in figure 120. If c is the capacitance value in the low pass circuit then the inductance and capacitance values in the band stop circuit are related to it as follows

$$C = \frac{c\Delta}{R\omega_0^2} \qquad\qquad L = \frac{R}{c\Delta} \qquad\qquad (165)$$

R in these transform equations is the source or load resistance in the band stop circuit.

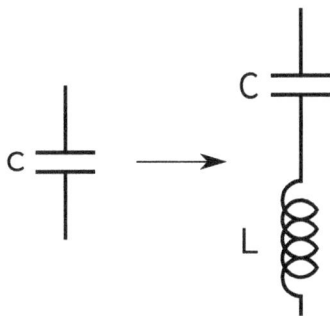

Figure 120: Shunt capacitor transformation.

Circuit diagrams for the second, fourth, sixth and eighth order band stop filters are shown on the following pages. The given component values will produce a band stop filter with $R = 1$, $\omega_0 = 1$, $\Delta = 1$, $\omega_1 = 1/\phi$ and $\omega_2 = \phi$ where $\phi = (\sqrt{5} + 1)/2$ is the golden ratio. The frequency response for these filters is shown in figure 121.

The component values in these circuits can easily be converted to other values of R, ω_0 and Δ by scaling the values in the series branches as follows

$$L \to \frac{LR\Delta}{\omega_0^2} \qquad\qquad C \to \frac{C}{R\Delta} \tag{166}$$

The values in the shunt branches are scaled as follows

$$C \to \frac{C\Delta}{R\omega_0^2} \qquad\qquad L \to \frac{LR}{\Delta} \tag{167}$$

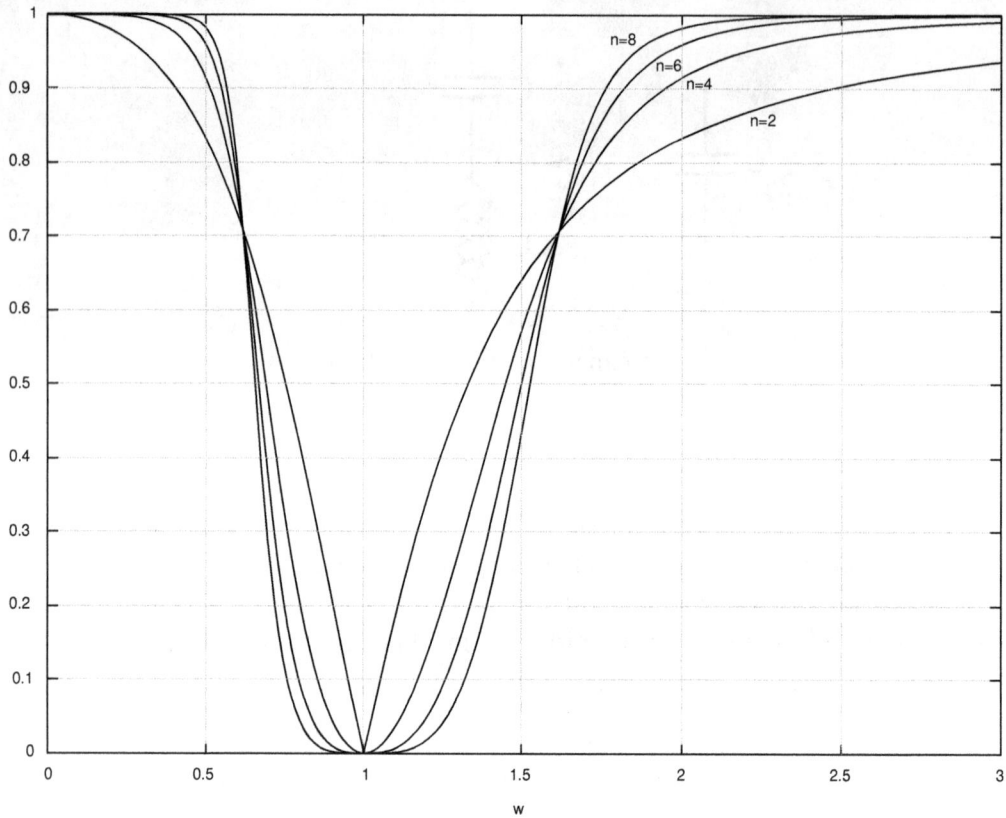

Figure 121: Frequency response for second, fourth, sixth and eighth order band stop filters.

For example if we want a fourth order band stop filter centered at 1 kHz with a bandwidth of 100 Hz and a source resistance of 50 Ω then using the component values in equation 173 we get, for $R = 50\Omega$, $\Delta = 2\pi \cdot 100$ and

$$\omega_0 = 2\pi \cdot 1000,$$

$$
\begin{aligned}
L_1 &\to L_1 R\Delta/\omega_0^2 & &\approx 563\mu\text{H} \\
C_1 &\to C_1/R\Delta & &\approx 45\mu\text{F} \\
C_2 &\to C_2\Delta/R\omega_0^2 & &\approx 0.45\mu\text{F} \\
L_2 &\to L_2 R/\Delta & &\approx 56\text{mH}
\end{aligned}
\tag{168}
$$

The upper and lower cutoff frequencies for the filter are calculated using eq. 163. They are $f_1 = 951.25$ Hz and $f_2 = 1051.25$ Hz.

The spice code and the frequency response simulation for the filter are shown below.

```
.title band stop 4th order with source resistance
V1 in 0 dc 0 ac 1
R1 in 1 50
L1 1 out 563u
C1 1 out 45u
C2 out 2 0.45u
L2 2 0 56m
.control
ac dec 1000 800 1200
gnuplot newplot vm(out)
.endc
.end
```

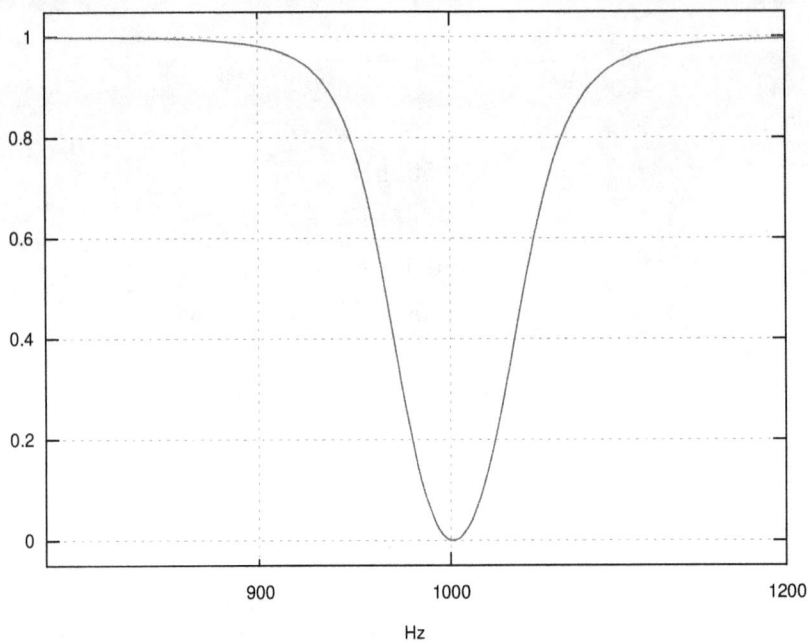

Figure 122: Spice generated frequency response for fourth order band stop filter with source resistance $R = 50\Omega$ and $L_1 = 563\mu H$, $C_1 = 45\mu F$, $C_2 = 0.45\mu F$ and $L_2 = 56\text{m}H$.

Second Order Band Stop Butterworth

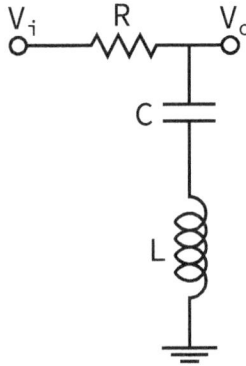

Figure 123: Second order band stop Butterworth with source impedance.

$$C = 1 \qquad\qquad L = 1 \qquad\qquad (169)$$

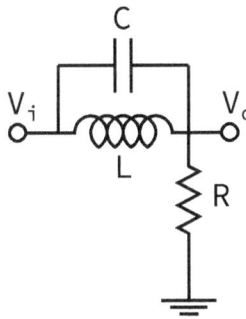

Figure 124: Second order band stop Butterworth with load impedance.

$$C = 1 \qquad\qquad L = 1 \qquad\qquad (170)$$

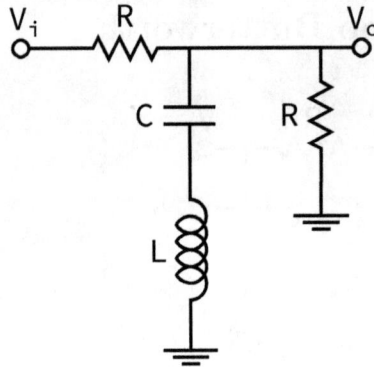

Figure 125: Second order band stop Butterworth with source and load impedance, variation 1.

$$C = 2 \qquad\qquad L = 1/2 \qquad\qquad (171)$$

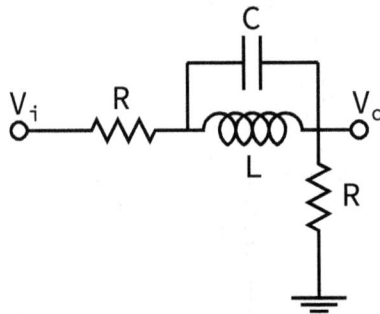

Figure 126: Second order band stop Butterworth with source and load impedance, variation 2.

$$C = 1/2 \qquad\qquad L = 2 \qquad\qquad (172)$$

Fourth Order Band Stop Butterworth

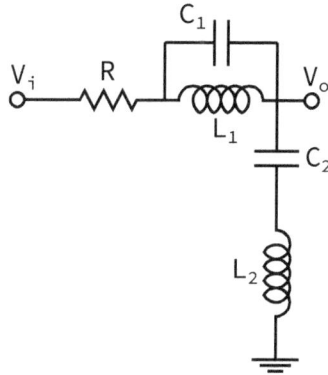

Figure 127: Fourth order band stop Butterworth with source impedance.

$$C_1 = \sqrt{2} \qquad L_1 = 1/\sqrt{2} \qquad C_2 = \sqrt{2} \qquad L_2 = 1/\sqrt{2} \qquad (173)$$

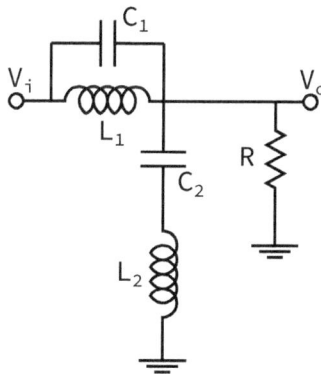

Figure 128: Fourth order band stop Butterworth with load impedance.

$$C_1 = 1/\sqrt{2} \qquad L_1 = \sqrt{2} \qquad C_2 = 1/\sqrt{2} \qquad L_2 = \sqrt{2} \qquad (174)$$

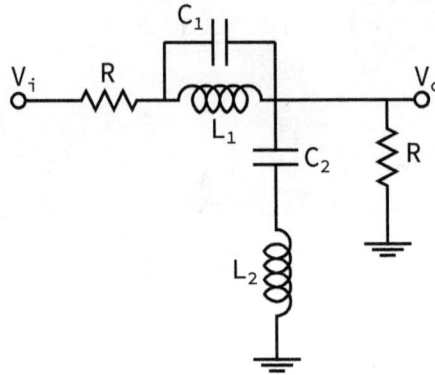

Figure 129: Fourth order band stop Butterworth with source and load impedance, variation 1.

$$C_1 = 1/\sqrt{2} \qquad L_1 = \sqrt{2} \qquad C_2 = \sqrt{2} \qquad L_2 = 1/\sqrt{2} \qquad (175)$$

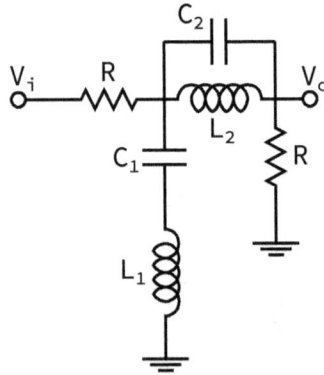

Figure 130: Fourth order band stop Butterworth with source and load impedance, variation 2.

$$C_1 = \sqrt{2} \qquad L_1 = 1/\sqrt{2} \qquad C_2 = 1/\sqrt{2} \qquad L_2 = \sqrt{2} \qquad (176)$$

Sixth Order Band Stop Butterworth

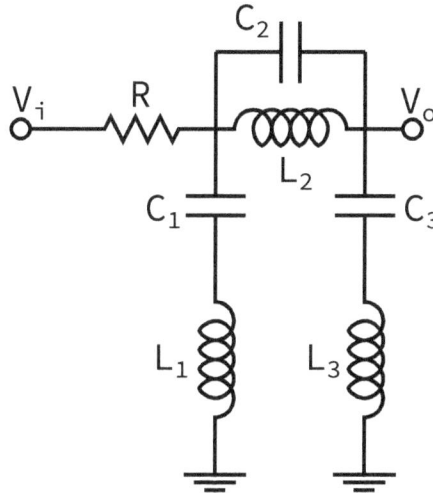

Figure 131: Sixth order band stop Butterworth with source impedance.

$C_1 = 1/2$ $L_1 = 2$

$C_2 = 3/4$ $L_2 = 4/3$

$C_3 = 3/2$ $L_3 = 2/3$ (177)

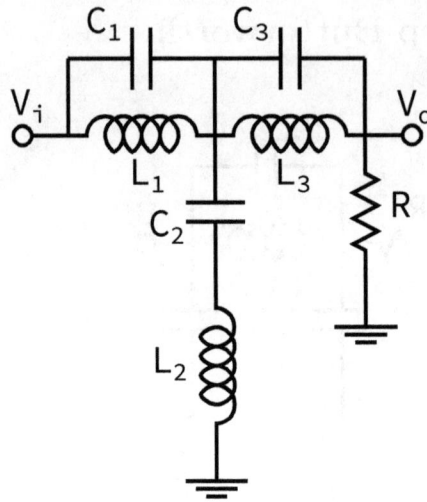

Figure 132: Sixth order band stop Butterworth with load impedance.

$$C_1 = 2/3 \qquad\qquad L_1 = 3/2$$
$$C_2 = 4/3 \qquad\qquad L_2 = 3/4$$
$$C_3 = 2 \qquad\qquad L_3 = 1/2 \qquad\qquad\qquad (178)$$

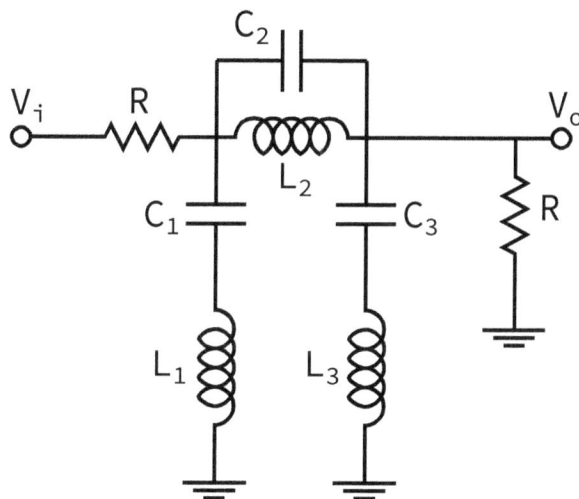

Figure 133: Sixth order band stop Butterworth with source and load impedance, variation 1.

$C_1 = 1$ $L_1 = 1$

$C_2 = 1/2$ $L_2 = 2$

$C_3 = 1$ $L_3 = 1$ (179)

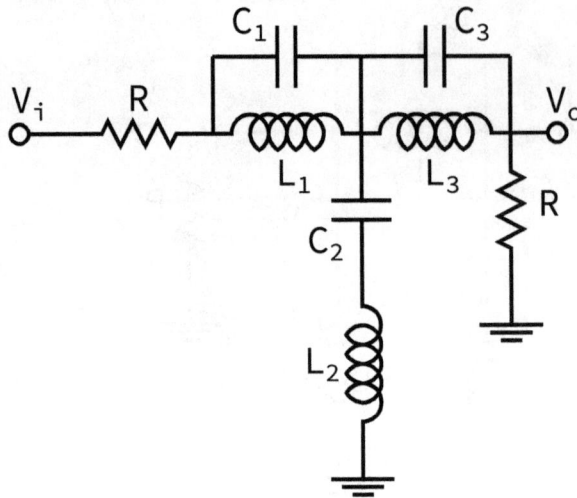

Figure 134: Sixth order band stop Butterworth with source and load impedance, variation 2.

$$C_1 = 1 \qquad\qquad L_1 = 1$$
$$C_2 = 2 \qquad\qquad L_2 = 1/2$$
$$C_3 = 1 \qquad\qquad L_3 = 1 \qquad\qquad\qquad (180)$$

Eighth Order Band Stop Butterworth

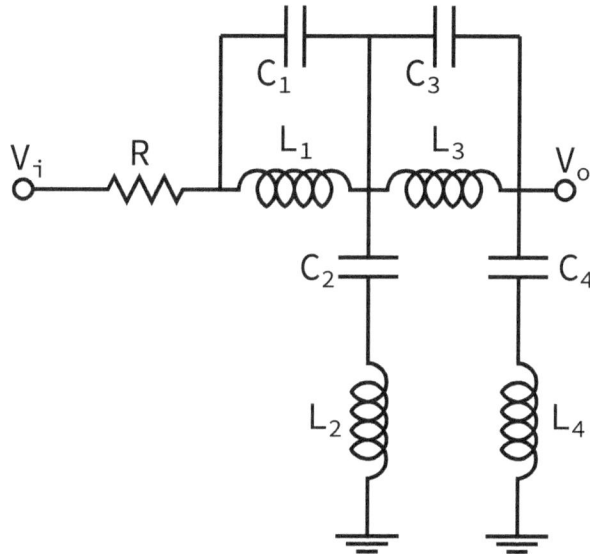

Figure 135: Eighth order band stop Butterworth with source impedance.

$C_1 = 2.6131259298$ $L_1 = 0.3826834324$

$C_2 = 1.0823922003$ $L_2 = 0.9238795325$

$C_3 = 0.6340506711$ $L_3 = 1.5771610149$

$C_4 = 1.5307337295$ $L_4 = 0.6532814824$ (181)

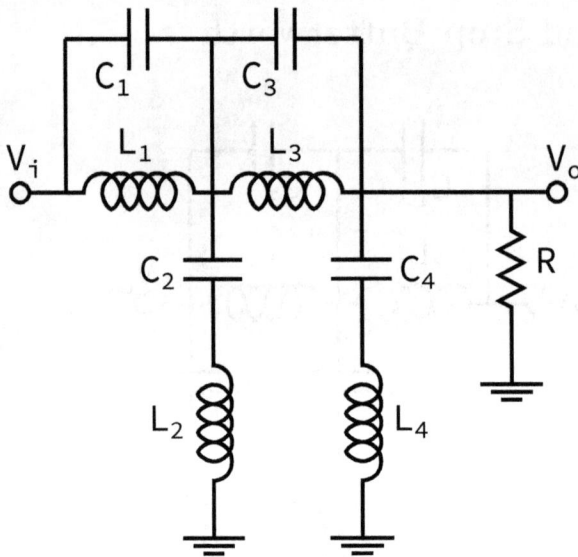

Figure 136: Eighth order band stop Butterworth with load impedance.

$$C_1 = 0.6532814824 \qquad L_1 = 1.5307337295$$
$$C_2 = 1.5771610149 \qquad L_2 = 0.6340506711$$
$$C_3 = 0.9238795325 \qquad L_3 = 1.0823922003$$
$$C_4 = 0.3826834324 \qquad L_4 = 2.6131259298 \tag{182}$$

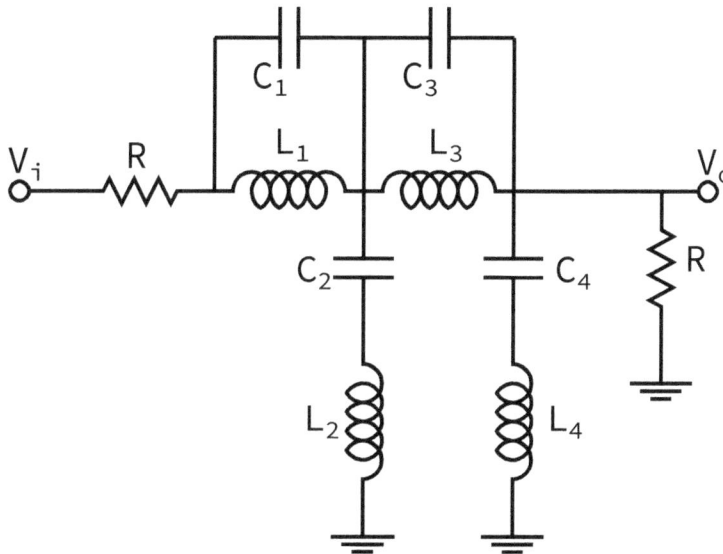

Figure 137: Eighth order band stop Butterworth with source and load impedance, variation 1.

$$C_1 = 1.3065629649 \qquad L_1 = 0.7653668647$$
$$C_2 = 1.8477590650 \qquad L_2 = 0.5411961001$$
$$C_3 = 0.5411961001 \qquad L_3 = 1.8477590650$$
$$C_4 = 0.7653668647 \qquad L_4 = 1.3065629649 \qquad (183)$$

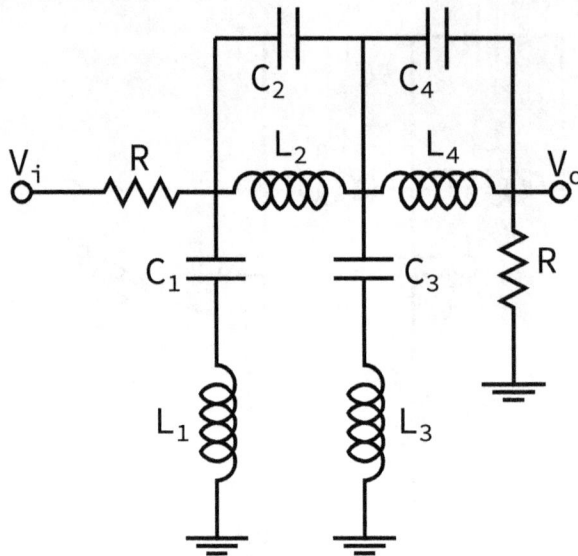

Figure 138: Eighth order band stop Butterworth with source and load impedance, variation 2.

$C_1 = 0.7653668647$

$C_2 = 0.5411961001$

$C_3 = 1.8477590650$

$C_4 = 1.3065629649$

$L_1 = 1.3065629649$

$L_2 = 1.8477590650$

$L_3 = 0.5411961001$

$L_4 = 0.7653668647$

(184)

REFERENCES & FURTHER READING

- *Passive and Active Network Analysis and Synthesis*, Aram Budak, 1974.
  http://www.worldcat.org/oclc/440086115

- *Analog Filter Design*, M.E. Van Valkenburg, 1982.
  http://www.worldcat.org/oclc/635208644
  This is the copy that we have. It's not easy to find, but the 2nd edition is, and is listed below:

- *Analog Filter Design*, Rolf Schaumann, Haiqiao Xiao, M.E. Van Valkenburg, 2nd ed, 2011.
  http://www.worldcat.org/oclc/656848106

- *Recursive Digital Filters: A Concise Guide*, Hollos and Hollos, 2014.
  http://www.abrazol.com/books/filter1/
  Contains digital Butterworth filters.

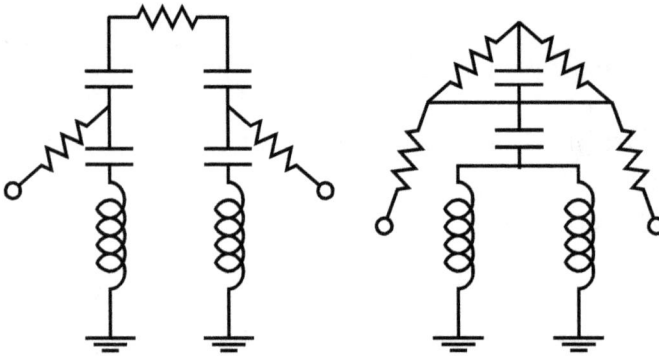

COMPONENTS

Resistor Standard Values

Table 1: Resistor standard values with 10% tolerance (E12).

1.0	10	100	1.0K	10K	100K	1.0M
1.2	12	120	1.2K	12K	120K	1.2M
1.5	15	150	1.5K	15K	150K	1.5M
1.8	18	180	1.8K	18K	180K	1.8M
2.2	22	220	2.2K	22K	220K	2.2M
2.7	27	270	2.7K	27K	270K	2.7M
3.3	33	330	3.3K	33K	330K	3.3M
3.9	39	390	3.9K	39K	390K	3.9M
4.7	47	470	4.7K	47K	470K	4.7M
5.6	56	560	5.6K	56K	560K	5.6M
6.8	68	680	6.8K	68K	680K	6.8M
8.2	82	820	8.2K	82K	820K	8.2M

Table 2: Resistor standard values with 5% tolerance (E24).

1.0	10	100	1.0K	10K	100K	1.0M
1.1	11	110	1.1K	11K	110K	1.1M
1.2	12	120	1.2K	12K	120K	1.2M
1.3	13	130	1.3K	13K	130K	1.3M
1.5	15	150	1.5K	15K	150K	1.5M
1.6	16	160	1.6K	16K	160K	1.6M
1.8	18	180	1.8K	18K	180K	1.8M
2.0	20	200	2.0K	20K	200K	2.0M
2.2	22	220	2.2K	22K	220K	2.2M
2.4	24	240	2.4K	24K	240K	2.4M
2.7	27	270	2.7K	27K	270K	2.7M
3.0	30	300	3.0K	30K	300K	3.0M
3.3	33	330	3.3K	33K	330K	3.3M
3.6	36	360	3.6K	36K	360K	3.6M
3.9	39	390	3.9K	39K	390K	3.9M
4.3	43	430	4.3K	43K	430K	4.3M
4.7	47	470	4.7K	47K	470K	4.7M
5.1	51	510	5.1K	51K	510K	5.1M
5.6	56	560	5.6K	56K	560K	5.6M
6.2	62	620	6.2K	62K	620K	6.2M
6.8	68	680	6.8K	68K	680K	6.8M
7.5	75	750	7.5K	75K	750K	7.5M
8.2	82	820	8.2K	82K	820K	8.2M
9.1	91	910	9.1K	91K	910K	9.1M

Table 3: Resistor standard values with 2% tolerance (E48).

1.00	10.0	100	1.00K	10.0K	100K	1.00M
1.05	10.5	105	1.05K	10.5K	105K	1.05M
1.10	11.0	110	1.10K	11.0K	110K	1.10M
1.15	11.5	115	1.15K	11.5K	115K	1.15M
1.21	12.1	121	1.21K	12.1K	121K	1.21M
1.27	12.7	127	1.27K	12.7K	127K	1.27M
1.33	13.3	133	1.33K	13.3K	133K	1.33M
1.40	14.0	140	1.40K	14.0K	140K	1.40M
1.47	14.7	147	1.47K	14.7K	147K	1.47M
1.54	15.4	154	1.54K	15.4K	154K	1.54M
1.62	16.2	162	1.62K	16.2K	162K	1.62M
1.69	16.9	169	1.69K	16.9K	169K	1.69M
1.78	17.8	178	1.78K	17.8K	178K	1.78M
1.87	18.7	187	1.87K	18.7K	187K	1.87M
1.96	19.6	196	1.96K	19.6K	196K	1.96M
2.05	20.5	205	2.05K	20.5K	205K	2.05M
2.15	21.5	215	2.15K	21.5K	215K	2.15M
2.26	22.6	226	2.26K	22.6K	226K	2.26M
2.37	23.7	237	2.37K	23.7K	237K	2.37M
2.49	24.9	249	2.49K	24.9K	249K	2.49M
2.61	26.1	261	2.61K	26.1K	261K	2.61M
2.74	27.4	274	2.74K	27.4K	274K	2.74M
2.87	28.7	287	2.87K	28.7K	287K	2.87M
3.01	30.1	301	3.01K	30.1K	301K	3.01M

continued...

Table 4: Resistor standard values with 2% tolerance (E48), continued.

3.16	31.6	316	3.16K	31.6K	316K	3.16M
3.32	33.2	332	3.32K	33.2K	332K	3.32M
3.48	34.8	348	3.48K	34.8K	348K	3.48M
3.65	36.5	365	3.65K	36.5K	365K	3.65M
3.83	38.3	383	3.83K	38.3K	383K	3.83M
4.02	40.2	402	4.02K	40.2K	402K	4.02M
4.22	42.2	422	4.22K	42.2K	422K	4.22M
4.42	44.2	442	4.42K	44.2K	442K	4.42M
4.64	46.4	464	4.64K	46.4K	464K	4.64M
4.87	48.7	487	4.87K	48.7K	487K	4.87M
5.11	51.1	511	5.11K	51.1K	511K	5.11M
5.36	53.6	536	5.36K	53.6K	536K	5.36M
5.62	56.2	562	5.62K	56.2K	562K	5.62M
5.90	59.0	590	5.90K	59.0K	590K	5.90M
6.19	61.9	619	6.19K	61.9K	619K	6.19M
6.49	64.9	649	6.49K	64.9K	649K	6.49M
6.81	68.1	681	6.81K	68.1K	681K	6.81M
7.15	71.5	715	7.15K	71.5K	715K	7.15M
7.50	75.0	750	7.50K	75.0K	750K	7.50M
7.87	78.7	787	7.87K	78.7K	787K	7.87M
8.25	82.5	825	8.25K	82.5K	825K	8.25M
8.66	86.6	866	8.66K	86.6K	866K	8.66M
9.09	90.9	909	9.09K	90.9K	909K	9.09M
9.53	95.3	953	9.53K	95.3K	953K	9.53M

Table 5: Resistor standard values with 1% tolerance (E96).

1.00	10.0	100	1.00K	10.0K	100K	1.00M
1.02	10.2	102	1.02K	10.2K	102K	1.02M
1.05	10.5	105	1.05K	10.5K	105K	1.05M
1.07	10.7	107	1.07K	10.7K	107K	1.07M
1.10	11.0	110	1.10K	11.0K	110K	1.10M
1.13	11.3	113	1.13K	11.3K	113K	1.13M
1.15	11.5	115	1.15K	11.5K	115K	1.15M
1.18	11.8	118	1.18K	11.8K	118K	1.18M
1.21	12.1	121	1.21K	12.1K	121K	1.21M
1.24	12.4	124	1.24K	12.4K	124K	1.24M
1.27	12.7	127	1.27K	12.7K	127K	1.27M
1.30	13.0	130	1.30K	13.0K	130K	1.30M
1.33	13.3	133	1.33K	13.3K	133K	1.33M
1.37	13.7	137	1.37K	13.7K	137K	1.37M
1.40	14.0	140	1.40K	14.0K	140K	1.40M
1.43	14.3	143	1.43K	14.3K	143K	1.43M
1.47	14.7	147	1.47K	14.7K	147K	1.47M
1.50	15.0	150	1.50K	15.0K	150K	1.50M
1.54	15.4	154	1.54K	15.4K	154K	1.54M
1.58	15.8	158	1.58K	15.8K	158K	1.58M
1.62	16.2	162	1.62K	16.2K	162K	1.62M
1.65	16.5	165	1.65K	16.5K	165K	1.65M
1.69	16.9	169	1.69K	16.9K	169K	1.69M
1.74	17.4	174	1.74K	17.4K	174K	1.74M
1.78	17.8	178	1.78K	17.8K	178K	1.78M
1.82	18.2	182	1.82K	18.2K	182K	1.82M
1.87	18.7	187	1.87K	18.7K	187K	1.87M
1.91	19.1	191	1.91K	19.1K	191K	1.91M
1.96	19.6	196	1.96K	19.6K	196K	1.96M
2.00	20.0	200	2.00K	20.0K	200K	2.00M
2.05	20.5	205	2.05K	20.5K	205K	2.05M
2.10	21.0	210	2.10K	21.0K	210K	2.10M

continued...

Table 6: Resistor standard values with 1% tolerance (E96), continued.

2.15	21.5	215	2.15K	21.5K	215K	2.15M
2.21	22.1	221	2.21K	22.1K	221K	2.21M
2.26	22.6	226	2.26K	22.6K	226K	2.26M
2.32	23.2	232	2.32K	23.2K	232K	2.32M
2.37	23.7	237	2.37K	23.7K	237K	2.37M
2.43	24.3	243	2.43K	24.3K	243K	2.43M
2.49	24.9	249	2.49K	24.9K	249K	2.49M
2.55	25.5	255	2.55K	25.5K	255K	2.55M
2.61	26.1	261	2.61K	26.1K	261K	2.61M
2.67	26.7	267	2.67K	26.7K	267K	2.67M
2.74	27.4	274	2.74K	27.4K	274K	2.74M
2.80	28.0	280	2.80K	28.0K	280K	2.80M
2.87	28.7	287	2.87K	28.7K	287K	2.87M
2.94	29.4	294	2.94K	29.4K	294K	2.94M
3.01	30.1	301	3.01K	30.1K	301K	3.01M
3.09	30.9	309	3.09K	30.9K	309K	3.09M
3.16	31.6	316	3.16K	31.6K	316K	3.16M
3.24	32.4	324	3.24K	32.4K	324K	3.24M
3.32	33.2	332	3.32K	33.2K	332K	3.32M
3.40	34.0	340	3.40K	34.0K	340K	3.40M
3.48	34.8	348	3.48K	34.8K	348K	3.48M
3.57	35.7	357	3.57K	35.7K	357K	3.57M
3.65	36.5	365	3.65K	36.5K	365K	3.65M
3.74	37.4	374	3.74K	37.4K	374K	3.74M
3.83	38.3	383	3.83K	38.3K	383K	3.83M
3.92	39.2	392	3.92K	39.2K	392K	3.92M
4.02	40.2	402	4.02K	40.2K	402K	4.02M
4.12	41.2	412	4.12K	41.2K	412K	4.12M
4.22	42.2	422	4.22K	42.2K	422K	4.22M
4.32	43.2	432	4.32K	43.2K	432K	4.32M
4.42	44.2	442	4.42K	44.2K	442K	4.42M
4.53	45.3	453	4.53K	45.3K	453K	4.53M

continued...

Table 7: Resistor standard values with 1% tolerance (E96), continued.

4.64	46.4	464	4.64K	46.4K	464K	4.64M
4.75	47.5	475	4.75K	47.5K	475K	4.75M
4.87	48.7	487	4.87K	48.7K	487K	4.87M
4.99	49.9	499	4.99K	49.9K	499K	4.99M
5.11	51.1	511	5.11K	51.1K	511K	5.11M
5.23	52.3	523	5.23K	52.3K	523K	5.23M
5.36	53.6	536	5.36K	53.6K	536K	5.36M
5.49	54.9	549	5.49K	54.9K	549K	5.49M
5.62	56.2	562	5.62K	56.2K	562K	5.62M
5.76	57.6	576	5.76K	57.6K	576K	5.76M
5.90	59.0	590	5.90K	59.0K	590K	5.90M
6.04	60.4	604	6.04K	60.4K	604K	6.04M
6.19	61.9	619	6.19K	61.9K	619K	6.19M
6.34	63.4	634	6.34K	63.4K	634K	6.34M
6.49	64.9	649	6.49K	64.9K	649K	6.49M
6.65	66.5	665	6.65K	66.5K	665K	6.65M
6.81	68.1	681	6.81K	68.1K	681K	6.81M
6.98	69.8	698	6.98K	69.8K	698K	6.98M
7.15	71.5	715	7.15K	71.5K	715K	7.15M
7.32	73.2	732	7.32K	73.2K	732K	7.32M
7.50	75.0	750	7.50K	75.0K	750K	7.50M
7.68	76.8	768	7.68K	76.8K	768K	7.68M
7.87	78.7	787	7.87K	78.7K	787K	7.87M
8.06	80.6	806	8.06K	80.6K	806K	8.06M
8.25	82.5	825	8.25K	82.5K	825K	8.25M
8.45	84.5	845	8.45K	84.5K	845K	8.45M
8.66	86.6	866	8.66K	86.6K	866K	8.66M
8.87	88.7	887	8.87K	88.7K	887K	8.87M
9.09	90.9	909	9.09K	90.9K	909K	9.09M
9.31	93.1	931	9.31K	93.1K	931K	9.31M
9.53	95.3	953	9.53K	95.3K	953K	9.53M
9.76	97.6	976	9.76K	97.6K	976K	9.76M

Capacitor Standard Values

Table 8: Capacitor standard values with 10% tolerance (E12).

10p	100p	1.0n	10n	100n	1.0u	10u
12p	120p	1.2n	12n	120n	1.2u	12u
15p	150p	1.5n	15n	150n	1.5u	15u
18p	180p	1.8n	18n	180n	1.8u	18u
22p	220p	2.2n	22n	220n	2.2u	22u
27p	270p	2.7n	27n	270n	2.7u	27u
33p	330p	3.3n	33n	330n	3.3u	33u
39p	390p	3.9n	39n	390n	3.9u	39u
47p	470p	4.7n	47n	470n	4.7u	47u
56p	560p	5.6n	56n	560n	5.6u	56u
68p	680p	6.8n	68n	680n	6.8u	68u
82p	820p	8.2n	82n	820n	8.2u	82u

Table 9: Capacitor standard values with 5% tolerance (E24).

10p	100p	1.0n	10n	100n	1.0u	10u
11p	110p	1.1n	11n	110n	1.1u	11u
12p	120p	1.2n	12n	120n	1.2u	12u
13p	130p	1.3n	13n	130n	1.3u	13u
15p	150p	1.5n	15n	150n	1.5u	15u
16p	160p	1.6n	16n	160n	1.6u	16u
18p	180p	1.8n	18n	180n	1.8u	18u
20p	200p	2.0n	20n	200n	2.0u	20u
22p	220p	2.2n	22n	220n	2.2u	22u
24p	240p	2.4n	24n	240n	2.4u	24u
27p	270p	2.7n	27n	270n	2.7u	27u
30p	300p	3.0n	30n	300n	3.0u	30u
33p	330p	3.3n	33n	330n	3.3u	33u
36p	360p	3.6n	36n	360n	3.6u	36u
39p	390p	3.9n	39n	390n	3.9u	39u
43p	430p	4.3n	43n	430n	4.3u	43u
47p	470p	4.7n	47n	470n	4.7u	47u
51p	510p	5.1n	51n	510n	5.1u	51u
56p	560p	5.6n	56n	560n	5.6u	56u
62p	620p	6.2n	62n	620n	6.2u	62u
68p	680p	6.8n	68n	680n	6.8u	68u
75p	750p	7.5n	75n	750n	7.5u	75u
82p	820p	8.2n	82n	820n	8.2u	82u
91p	910p	9.1n	91n	910n	9.1u	91u

Table 10: Capacitor standard values with 2% tolerance (E48).

10.0p	100p	1.00n	10.0n	100n	1.00u	10.0u
10.5p	105p	1.05n	10.5n	105n	1.05u	10.5u
11.0p	110p	1.10n	11.0n	110n	1.10u	11.0u
11.5p	115p	1.15n	11.5n	115n	1.15u	11.5u
12.1p	121p	1.21n	12.1n	121n	1.21u	12.1u
12.7p	127p	1.27n	12.7n	127n	1.27u	12.7u
13.3p	133p	1.33n	13.3n	133n	1.33u	13.3u
14.0p	140p	1.40n	14.0n	140n	1.40u	14.0u
14.7p	147p	1.47n	14.7n	147n	1.47u	14.7u
15.4p	154p	1.54n	15.4n	154n	1.54u	15.4u
16.2p	162p	1.62n	16.2n	162n	1.62u	16.2u
16.9p	169p	1.69n	16.9n	169n	1.69u	16.9u
17.8p	178p	1.78n	17.8n	178n	1.78u	17.8u
18.7p	187p	1.87n	18.7n	187n	1.87u	18.7u
19.6p	196p	1.96n	19.6n	196n	1.96u	19.6u
20.5p	205p	2.05n	20.5n	205n	2.05u	20.5u
21.5p	215p	2.15n	21.5n	215n	2.15u	21.5u
22.6p	226p	2.26n	22.6n	226n	2.26u	22.6u
23.7p	237p	2.37n	23.7n	237n	2.37u	23.7u
24.9p	249p	2.49n	24.9n	249n	2.49u	24.9u
26.1p	261p	2.61n	26.1n	261n	2.61u	26.1u
27.4p	274p	2.74n	27.4n	274n	2.74u	27.4u
28.7p	287p	2.87n	28.7n	287n	2.87u	28.7u
30.1p	301p	3.01n	30.1n	301n	3.01u	30.1u

continued...

Table 11: Capacitor standard values with 2% tolerance (E48), continued.

31.6p	316p	3.16n	31.6n	316n	3.16u	31.6u
33.2p	332p	3.32n	33.2n	332n	3.32u	33.2u
34.8p	348p	3.48n	34.8n	348n	3.48u	34.8u
36.5p	365p	3.65n	36.5n	365n	3.65u	36.5u
38.3p	383p	3.83n	38.3n	383n	3.83u	38.3u
40.2p	402p	4.02n	40.2n	402n	4.02u	40.2u
42.2p	422p	4.22n	42.2n	422n	4.22u	42.2u
44.2p	442p	4.42n	44.2n	442n	4.42u	44.2u
46.4p	464p	4.64n	46.4n	464n	4.64u	46.4u
48.7p	487p	4.87n	48.7n	487n	4.87u	48.7u
51.1p	511p	5.11n	51.1n	511n	5.11u	51.1u
53.6p	536p	5.36n	53.6n	536n	5.36u	53.6u
56.2p	562p	5.62n	56.2n	562n	5.62u	56.2u
59.0p	590p	5.90n	59.0n	590n	5.90u	59.0u
61.9p	619p	6.19n	61.9n	619n	6.19u	61.9u
64.9p	649p	6.49n	64.9n	649n	6.49u	64.9u
68.1p	681p	6.81n	68.1n	681n	6.81u	68.1u
71.5p	715p	7.15n	71.5n	715n	7.15u	71.5u
75.0p	750p	7.50n	75.0n	750n	7.50u	75.0u
78.7p	787p	7.87n	78.7n	787n	7.87u	78.7u
82.5p	825p	8.25n	82.5n	825n	8.25u	82.5u
86.6p	866p	8.66n	86.6n	866n	8.66u	86.6u
90.9p	909p	9.09n	90.9n	909n	9.09u	90.9u
95.3p	953p	9.53n	95.3n	953n	9.53u	95.3u

Table 12: Capacitor standard values with 1% tolerance (E96).

10.0p	100p	1.00n	10.0n	100n	1.00u	10.0u
10.2p	102p	1.02n	10.2n	102n	1.02u	10.2u
10.5p	105p	1.05n	10.5n	105n	1.05u	10.5u
10.7p	107p	1.07n	10.7n	107n	1.07u	10.7u
11.0p	110p	1.10n	11.0n	110n	1.10u	11.0u
11.3p	113p	1.13n	11.3n	113n	1.13u	11.3u
11.5p	115p	1.15n	11.5n	115n	1.15u	11.5u
11.8p	118p	1.18n	11.8n	118n	1.18u	11.8u
12.1p	121p	1.21n	12.1n	121n	1.21u	12.1u
12.4p	124p	1.24n	12.4n	124n	1.24u	12.4u
12.7p	127p	1.27n	12.7n	127n	1.27u	12.7u
13.0p	130p	1.30n	13.0n	130n	1.30u	13.0u
13.3p	133p	1.33n	13.3n	133n	1.33u	13.3u
13.7p	137p	1.37n	13.7n	137n	1.37u	13.7u
14.0p	140p	1.40n	14.0n	140n	1.40u	14.0u
14.3p	143p	1.43n	14.3n	143n	1.43u	14.3u
14.7p	147p	1.47n	14.7n	147n	1.47u	14.7u
15.0p	150p	1.50n	15.0n	150n	1.50u	15.0u
15.4p	154p	1.54n	15.4n	154n	1.54u	15.4u
15.8p	158p	1.58n	15.8n	158n	1.58u	15.8u
16.2p	162p	1.62n	16.2n	162n	1.62u	16.2u
16.5p	165p	1.65n	16.5n	165n	1.65u	16.5u
16.9p	169p	1.69n	16.9n	169n	1.69u	16.9u
17.4p	174p	1.74n	17.4n	174n	1.74u	17.4u
17.8p	178p	1.78n	17.8n	178n	1.78u	17.8u
18.2p	182p	1.82n	18.2n	182n	1.82u	18.2u
18.7p	187p	1.87n	18.7n	187n	1.87u	18.7u
19.1p	191p	1.91n	19.1n	191n	1.91u	19.1u
19.6p	196p	1.96n	19.6n	196n	1.96u	19.6u
20.0p	200p	2.00n	20.0n	200n	2.00u	20.0u
20.5p	205p	2.05n	20.5n	205n	2.05u	20.5u
21.0p	210p	2.10n	21.0n	210n	2.10u	21.0u

continued...

Table 13: Capacitor standard values with 1% tolerance (E96), continued.

21.5p	215p	2.15n	21.5n	215n	2.15u	21.5u
22.1p	221p	2.21n	22.1n	221n	2.21u	22.1u
22.6p	226p	2.26n	22.6n	226n	2.26u	22.6u
23.2p	232p	2.32n	23.2n	232n	2.32u	23.2u
23.7p	237p	2.37n	23.7n	237n	2.37u	23.7u
24.3p	243p	2.43n	24.3n	243n	2.43u	24.3u
24.9p	249p	2.49n	24.9n	249n	2.49u	24.9u
25.5p	255p	2.55n	25.5n	255n	2.55u	25.5u
26.1p	261p	2.61n	26.1n	261n	2.61u	26.1u
26.7p	267p	2.67n	26.7n	267n	2.67u	26.7u
27.4p	274p	2.74n	27.4n	274n	2.74u	27.4u
28.0p	280p	2.80n	28.0n	280n	2.80u	28.0u
28.7p	287p	2.87n	28.7n	287n	2.87u	28.7u
29.4p	294p	2.94n	29.4n	294n	2.94u	29.4u
30.1p	301p	3.01n	30.1n	301n	3.01u	30.1u
30.9p	309p	3.09n	30.9n	309n	3.09u	30.9u
31.6p	316p	3.16n	31.6n	316n	3.16u	31.6u
32.4p	324p	3.24n	32.4n	324n	3.24u	32.4u
33.2p	332p	3.32n	33.2n	332n	3.32u	33.2u
34.0p	340p	3.40n	34.0n	340n	3.40u	34.0u
34.8p	348p	3.48n	34.8n	348n	3.48u	34.8u
35.7p	357p	3.57n	35.7n	357n	3.57u	35.7u
36.5p	365p	3.65n	36.5n	365n	3.65u	36.5u
37.4p	374p	3.74n	37.4n	374n	3.74u	37.4u
38.3p	383p	3.83n	38.3n	383n	3.83u	38.3u
39.2p	392p	3.92n	39.2n	392n	3.92u	39.2u
40.2p	402p	4.02n	40.2n	402n	4.02u	40.2u
41.2p	412p	4.12n	41.2n	412n	4.12u	41.2u
42.2p	422p	4.22n	42.2n	422n	4.22u	42.2u
43.2p	432p	4.32n	43.2n	432n	4.32u	43.2u
44.2p	442p	4.42n	44.2n	442n	4.42u	44.2u
45.3p	453p	4.53n	45.3n	453n	4.53u	45.3u

continued...

Table 14: Capacitor standard values with 1% tolerance (E96), continued.

46.4p	464p	4.64n	46.4n	464n	4.64u	46.4u
47.5p	475p	4.75n	47.5n	475n	4.75u	47.5u
48.7p	487p	4.87n	48.7n	487n	4.87u	48.7u
49.9p	499p	4.99n	49.9n	499n	4.99u	49.9u
51.1p	511p	5.11n	51.1n	511n	5.11u	51.1u
52.3p	523p	5.23n	52.3n	523n	5.23u	52.3u
53.6p	536p	5.36n	53.6n	536n	5.36u	53.6u
54.9p	549p	5.49n	54.9n	549n	5.49u	54.9u
56.2p	562p	5.62n	56.2n	562n	5.62u	56.2u
57.6p	576p	5.76n	57.6n	576n	5.76u	57.6u
59.0p	590p	5.90n	59.0n	590n	5.90u	59.0u
60.4p	604p	6.04n	60.4n	604n	6.04u	60.4u
61.9p	619p	6.19n	61.9n	619n	6.19u	61.9u
63.4p	634p	6.34n	63.4n	634n	6.34u	63.4u
64.9p	649p	6.49n	64.9n	649n	6.49u	64.9u
66.5p	665p	6.65n	66.5n	665n	6.65u	66.5u
68.1p	681p	6.81n	68.1n	681n	6.81u	68.1u
69.8p	698p	6.98n	69.8n	698n	6.98u	69.8u
71.5p	715p	7.15n	71.5n	715n	7.15u	71.5u
73.2p	732p	7.32n	73.2n	732n	7.32u	73.2u
75.0p	750p	7.50n	75.0n	750n	7.50u	75.0u
76.8p	768p	7.68n	76.8n	768n	7.68u	76.8u
78.7p	787p	7.87n	78.7n	787n	7.87u	78.7u
80.6p	806p	8.06n	80.6n	806n	8.06u	80.6u
82.5p	825p	8.25n	82.5n	825n	8.25u	82.5u
84.5p	845p	8.45n	84.5n	845n	8.45u	84.5u
86.6p	866p	8.66n	86.6n	866n	8.66u	86.6u
88.7p	887p	8.87n	88.7n	887n	8.87u	88.7u
90.9p	909p	9.09n	90.9n	909n	9.09u	90.9u
93.1p	931p	9.31n	93.1n	931n	9.31u	93.1u
95.3p	953p	9.53n	95.3n	953n	9.53u	95.3u
97.6p	976p	9.76n	97.6n	976n	9.76u	97.6u

Resistor Color Codes

Table 15: Resistor Color Codes.

Color	Value	Multiplier	Tolerance
Black	0	1	
Brown	1	10	$\pm 1\%$
Red	2	10^2	$\pm 2\%$
Orange	3	10^3	
Yellow	4	10^4	
Green	5	10^5	
Blue	6	10^6	
Violet	7	10^7	
Gray	8	10^8	
White	9	10^9	
Gold		10^{-1}	$\pm 5\%$
Silver		10^{-2}	$\pm 10\%$
Pink		10^{-3}	

Capacitor Codes

If a capacitor has only a two digit number then that is the value in pico-farads. If it has a three digit number then the last digit is the power of 10 multiplier. Some examples are below.

$104 \rightarrow 0.1\mu F = 100nF$
$103 \rightarrow 0.01\mu F = 10nF$
$102 \rightarrow 0.001\mu F = 1nF$
$101 \rightarrow 0.0001\mu F = 100pF$
$100 \rightarrow 10pF$

$220 \rightarrow 22\text{pF}$
$330 \rightarrow 33\text{pF}$
$470 \rightarrow 47\text{pF}$

Capacitor tolerance is designated by a letter following the number. The letters are as follows:

D 0.5%
F 1%
G 2%
H 3%
J 5%
K 10%
M 20%
Z +80/-20%

Inductor Manufacturers

Inductors are generally not hard to make yourself. Here is a nice resource for doing that:

Homebrew Your Own Inductors!
http://www.arrl.org/files/file/Technology/tis/info/pdf/9708033.pdf

If you want to buy your inductors, some companies that make them are:

Coilcraft	https://www.coilcraft.com/
muRata	https://www.murata.com/en-us/products/inductor
Vishay	https://www.vishay.com/inductors/
TDK	https://www.tdk-electronics.tdk.com/en/inductors
AVX	https://www.avx.com/products/inductors/
Abracon	https://abracon.com/parametric/inductors
West Coast Magnetics	https://www.wcmagnetics.com/inductors/

ACKNOWLEDGMENTS

In ordinary life we hardly realize that we receive a great deal more than we give, and that it is only with gratitude that life becomes rich. It is very easy to overestimate the importance of our own achievements in comparison with what we owe to others.

Dietrich Bonhoeffer, letter to parents from prison, Sept. 13, 1943

We'd like to thank our parents, Istvan and Anna Hollos, for helping us in many ways.

We thank the makers and maintainers of all the software we've used in the production of this book, including: gcc, Emacs text editor, LaTeX typesetting system, Inkscape, Ngspice, mupdf and evince document viewers, bash shell, and the GNU/Linux operating system.

ABOUT THE AUTHORS

Stefan Hollos and **J. Richard Hollos** are physicists and electrical engineers by training, and enjoy anything related to math, physics, engineering and computing. In addition, they enjoy creating music and visual art, and being in the great outdoors. They are the authors of:

- Nell: An SVG Drawing Language

- Coin Tossing: The Hydrogen Atom of Probability

- Creating Melodies

- Hexagonal Tilings and Patterns

- Combinatorics II Problems and Solutions: Counting Patterns

- Information Theory: A Concise Introduction

- Recursive Digital Filters: A Concise Guide

- Art of Pi

- Creating Noise

- Art of the Golden Ratio

- Creating Rhythms

- Pattern Generation for Computational Art

- Finite Automata and Regular Expressions: Problems and Solutions

- **Probability Problems and Solutions**

- **Combinatorics Problems and Solutions**

- **The Coin Toss: Probabilities and Patterns**

- **Pairs Trading: A Bayesian Example**

- **Simple Trading Strategies That Work**

- **Bet Smart: The Kelly System for Gambling and Investing**

- **Signals from the Subatomic World: How to Build a Proton Precession Magnetometer**

They are brothers and business partners at Exstrom Laboratories LLC in Longmont, Colorado. Their website is exstrom.com

THANK YOU

Thank you for buying this book.

If you'd like to receive news about this book and others published by Abrazol Publishing, just go to

http://www.abrazol.com/

and sign up for our newsletter.